TIME AND THE PSYCHE

In *Time and the Psyche*, a diverse selection of contributors explores the multi-layered aspects of time through the lens of analytical psychology. The book aims to bridge the gap between theory and practice, emphasising time's fundamental role in the workings and expressions of the psyche, and additionally exploring cultural and clinical dimensions.

The contributors deal with temporality in our inner world and its manifestations as expressed by products of our psyche, covering topics such as disturbances of temporality within the psychoanalytic session, the acausal connecting principle of synchronicity, time as expressed in film, objects, literature, and culture, and temporality as understood in various types of dreams and imaginary practices. The book also explores the time-bound world, time versus timelessness, the realm of the eternal, human versus cosmic time, Chronos versus Kairos, and other temporality-related dimensions and their relationship to our psyche and our experience in the world. With contributors from backgrounds in clinical work, the arts, literature, and philosophy, this collection is unique in its scope.

Time and the Psyche is a thought-provoking reading for academics and students of Jungian and post-Jungian studies, analytical psychologists, and Jungian analysts in practice and in training.

Angeliki Yiassemides, PhD, is an independent Jungian scholar, a developmental psychologist, and a Jungian analyst in training. She holds a PhD in psychoanalytic studies and is the author of *Time and Timelessness: Temporality in the Theory of Carl Jung* (Routledge).

'You are fortunate to have in your hands an exciting, engaging, and wonderfully edited volume stemming from a conference on temporality in analysis. Angeliki Yassemides has rendered a great service to analytical psychology in putting together the conference and this subsequent compilation of papers that plumb the depths of time, especially as experienced in psychological reality. Multifaceted explorations of clinical, cultural, and historical reflections on time offer readers the opportunity for a plunge into the complexity we call time. Do take the time to savor this delightful collection.'

Joe Cambray, Ph.D,
Provost, Pacifica Graduate Institute; Past-President, IAAP

'Angeliki Yiassemides has assembled a richly stocked treasure chest of a book. Each individual essay is a gem, and the totality amounts to a brilliant display of psychological insight into a multitude of experiences of temporality. This work is bound to generate much further discussion and interest in a topic of contemporary interest for all schools of psychotherapy.'

Murray Stein,
Author of *Minding the Self*

'The counting and accounting of time, which passes at its own rhythms, the individual and collective feeling of duration, the experience of living in different temporalities, some of them simultaneous, the amazement we feel when the past, supposedly gone forever, proves to be present still, or abruptly returns, or stubbornly repeats itself: all of these events occur daily. They continually mobilize our attention and wonderment, whether we are farming, strolling, making art, or researching physics or philosophy.

This age-old amazement and exploration come to life again, renewed by depth psychology, in various contemporary psychoanalytical traditions, and particularly in the works of Jung and of today's Jungian analysts and researchers. Recently, Angeliki Yiassemides reopened the proceedings masterfully, with *Time and Timelessness: Temporality in the Theory of Carl Jung* (Routledge, 2014). In this new book, she brings together a set of authors which includes some of the most influential thinkers in Jungian psychology and psychoanalysis. Within the scope of his own research, each essayist reconsiders the investigations and theories elaborated so far on the subject of time and temporality, suggesting new approaches and opening perspectives to help us recognize and think about time in the diversity of its expressions and dimensions.

I fervently recommend this book to every curious mind, to researchers, scholars, and clinicians. It provides abundant food for thought and observation.'

Christian Gaillard
Former Professor at the National Academy of Fine Arts,
Paris; Past-President, IAAP

TIME AND THE PSYCHE

Jungian Perspectives

Edited by Angeliki Yiassemides

Routledge
Taylor & Francis Group
LONDON AND NEW YORK

First published 2017
by Routledge
2 Park Square, Milton Park, Abingdon, Oxon OX14 4RN

and by Routledge
711 Third Avenue, New York, NY 10017

Routledge is an imprint of the Taylor & Francis Group, an informa business

British Library Cataloguing in Publication Data
A catalogue record for this book is available from the British Library

Library of Congress Cataloging in Publication Data
Names: Yiassemides, Angeliki, editor.
Title: Time and the psyche : Jungian perspectives / edited by Angeliki Yiassemides.
Description: Abingdon, Oxon ; New York, NY : Routledge, 2017. | Includes bibliographical references.
Identifiers: LCCN 2016049345| ISBN 9781138120716 (hardback : alk. paper) | ISBN 9781138120723 (pbk. : alk. paper) | ISBN 9781315651569 (ebook)
Subjects: | MESH: Time | Jungian Theory
Classification: LCC BF109.J8 | NLM WM 460.5.J9 |
DDC 155.2/644—dc23
LC record available at https://lccn.loc.gov/2016049345

ISBN: 978-1-138-12071-6 (hbk)
ISBN: 978-1-138-12072-3 (pbk)
ISBN: 978-1-315-65156-9 (ebk)

Typeset in Bembo
by Florence Production Ltd, Stoodleigh, Devon, UK

To my mother, Maritsa, for her time,
without which I would have none of my own.

Time is the substance I am made of. Time is a river that grabs me, but I am the river. It is a tiger that rends me, but I am the tiger. It is a fire that consumes me, but I am the fire.

(Borges 1974)

CONTENTS

FIGURES

CONTRIBUTORS

Paul Bishop is William Jacks Professor of Modern Languages at the University of Glasgow, where he has taught since 1994. His major publications include *Analytical Psychology and German Classical Aesthetics*, 2 vols (2008–2009), *Reading Goethe at Midlife* (2011) and *On the Blissful Islands* (forthcoming). He is editor of *Jung in Contexts: A Reader* (1999) and *The Archaic: The Past in the Present* (2012).

Laner Cassar, PhD, is a clinical psychologist and psychotherapist, who holds a doctorate degree in psychoanalytic studies at the University of Essex. He is currently completing his Jungian training with CIPA, Southern Institute, Italy and the International Association of Analytical Psychology (IAAP). He is founder and president of the Malta Depth Psychological Association. His publications and projects include *Robert Desoille – A Master of the Imagination* (2014, video documentary), *Shades of Red: Jung's Technique in the Red Book and Desoille's R.E.D (Rêve éveillé dirigé) Method* (2014) and *Words as Images in Gaston Bachelard and Robert Desoille* (in press).

Angela Connolly, MD, is a psychiatrist and Jungian analyst in Rome. She is a trainer, faculty member and supervisor of Centro Italiano di Psicologia Analitica (CIPA). She served as deputy editor of the *Journal of Analytical Psychology* (JAP) and is currently on the advisory board. Connolly was the Honorary Secretary of the IAAP (2010–2013) and she is currently serving as the Vice-President. She has published in English and Italian and has had articles translated in Russian and German. Her publications include: *Cognitive aesthetics of alchemical imagery* (JAP), *The cognitive aesthetics of dream metaphors* (in *Understanding Jung*, edited by A. Iapoce) and *Out of the body: embodiment and its vicissitudes* (JAP).

Christopher Hauke, PhD, is a Jungian analyst, Senior Lecturer at Goldsmiths, University of London and a consultant for film and digital media. His publications include: *Jung and the Postmodern. The Interpretation of Realities*; *Human Being Human. Culture and the Soul* and his latest book *Visible Mind: Movies, Modernity and the Unconscious*. He has co-edited two collections of Jungian writing on the moving image titled *Jung and Film: Post-Jungian Takes on the Moving Image* and *Jung and Film Two: The Return*. Christopher's own films have premiered in London, Barcelona and Montreal and he is the Jungian psychology consultant for Portal Entertainment, overseeing script and story development, script assessment, and narrative structuring for multi-platform, interactive and immersive productions.

Ladson Hinton, MD, trained in psychiatry at Stanford, and is a graduate of the C. G. Jung Institute of San Francisco. He is a member of the Society of Jungian Analysts of Northern California, the Institute of Contemporary Psychoanalysis in Los Angeles, and is a founding member of the New School for Analytical Psychology in Seattle. He is on the editorial board of the *Journal of Analytical Psychology* and practices, consults and teaches in Seattle. His recent publications are in the area of temporality and the philosophical and historical grounds of psychoanalysis.

J. A. Swan trained formally as a painter, and later qualified as a psychologist in clinical and transpersonal modalities. She holds a research PhD from the Centre for Psycholanalytic Studies in England, and is the author of *On the Spirit and the Self: The Religious Art of Marc Chagall* (Chiron, 2017). Her research interests combine fine arts and architecture, religious studies, and Jungian theory. She lives in Cyprus where she works as a writer and visual artist.

Caterina Vezzoli, PhD, is a clinical psychologist trained at the C. G. Jung Institute in Zurich, where she is serving as a training analyst. She is a training analyst, supervisor and faculty member at CIPA Southern Institute. Dr Vezzoli is the author and co-author of numerous scientific papers and books including: *Methodological evolution and clinical application of C. G. Jung's Word Association Experiment: a follow-up study* (Journal of Analytical Psychology 52(1)) and *Jung Today: Childhood and Adolescence* (Psychology Research Progress Series vol. 1).

Angeliki Yiassemides, PhD, is an independent Jungian scholar and a Developmental Psychologist (Columbia University, in the City of New York). Dr Yiassemides has a particular, long-standing interest in temporality in psychoanalytic theory and practice. Her time-related publications include *Chronos in synchronicity: manifestations of the psychoid reality* (*Journal of Analytical Psychology* 56(4)) and *Time and Timelessness: Temporality in the Theory of Carl Jung* (Routledge 2014/2016). She is currently in training with Centro Italiano di Psicologia Analytica (CIPA, Southern Institute, Italy) and the International Association of Analytical Psychology (IAAP). She lives and works in Nicosia, Cyprus.

EDITOR'S FOREWORD

Time and the Psyche: two incredibly obscure concepts paired to form the title of this book, adding, perhaps, to each other's ambiguity. How does time unfold in the psyche? What is its underlying reality within, and what are its manifestations in our endeavours? Are we clocks? Are we clock-makers? Time observers? Are we Time? *Time and the Psyche* attempts to approach these, and many more, questions through the lens of Analytical Psychology: It explores the ways in which time manifests and expresses in the psyche from a Jungian perspective, and how it relates with the overall opus of Jung's theory.

It is surprising that this volume is the first compilation of chapters that deal with temporality from a Jungian perspective, given the fact that the concept of time has occupied Carl Jung quite extensively. The most notable event which focused on the concept of Time that happened within Jung's lifetime and to which he contributed – even though it was not exclusively 'Jungian' in nature – took place in 1951 at the home of Frau Olga Froebe-Kapteyn, on the shore of Lake Maggiore in Switzerland: That year's *Eranos* meeting was devoted to *Man and Time.*

Joseph Campbell, who edited that third volume of the *Eranos Yearbooks*, wrote: 'The sureties were particularly well pulverized in the meeting of 1951, which was devoted to the mystery of Man and Time: a subject nicely calculated to make difficult the maintenance of any footing whatsoever' (Campbell 1957: xi). This is a timeless observation regarding 'the mystery of man and time'. It recalls St. Augustine's words about the nature of time, written in AD 397–398, which are among the most often-quoted musings, an aphorism indeed about time: 'What then is time? Provided that no one asks me, I know. If I want to explain it to an inquirer, I do not know' (Saint Augustine 1998: 230). And yet, we want to know.

Time and the Psyche is an exploration of temporality – its manifestations, meaning and magnitude. This, I believe, is a journey that each one of us must embark upon; it is an inevitable journey:

> There is always an implicit (if not explicit) assumption regarding time behind almost every theory concerning the nature of human existence. Each field, approach or statement needs to rest against a temporal background; each uttered sentence assumes a temporal frame. It is as if time is the barometer for every human expression.
>
> (Yiassemides 2014/2016: xiii)

The one requirement that is present in every language (at least the documented ones) is a verb. And each verb has a 'tense' – which in the Greek language is called 'chronos' (*χρόνος*); each verb has a chronos, that is, *a temporal hypostasis*. Each verb – which is the necessary ingredient for any sentence, which in turn is the building block of language, of communication and thinking – has temporality embedded in it.

Chronos and Kairos: temporal oscillations

Chronos, the personification of time for the Ancient Greeks, was one of the creative forces of the world: He 'produced from within himself the calm Aether and likewise Chaos, the empty space which had no firm bottom and was filled with darkness' (Kerényi 2006: 114). There is a common misconception that Chronos and Kronos, the Titan, are the same entity. This is perhaps due to the similarity (in English) of their names. Apart from this linguistic misunderstanding, there is a connection in their genealogy: Chronos gave birth to the hermaphroditic god Phanes, who created Ouranos (Heaven) and Gaia (Earth), who were the parents of the Titans, including Kronos. Therefore, Chronos is the great-grandfather of Kronos. (Not a very surprising turn of events, considering Ancient Greek lineage.) I have suggested elsewhere that 'this misunderstanding could . . . be due to the fact that the terrible qualities of Kronos happen to match our perception of time: it devours us; its passage causes our annihilation' (Yiassemides 2014/2016: xiv). Chronos, which assumes a linear unfolding of time, is intimately associated with death. As *Seíkilos* scripted upon the grave of his beloved Euterpe 'Life exists only for a short while; and Chronos demands an end.'[1]

Chronos is associated, and embedded (literally, i.e. etymologically), in chronological time: with the sequential passage of time, time's directionality towards the future, its discrete demarcations and their succession. The *lived experience* of time has been attributed to another entity, Kairos,[2] who is often positioned against Chronos in the epistemological analysis of temporality.

Chronos and Kairos are two of the protagonists of this volume; even when not mentioned explicitly, these pages are steeped with their respective temporalities.

As Laner Cassar explains, 'The Greeks . . . used the word *kairos* to describe the fullness of a sacred moment-of-truth or revelation and in this sense of time and space it captures the powerful impact of facing one's unconscious' [p. 57]. Kairos has been identified with the *opportune moment*, the (brief) moment in time in which things are possible. This implies a supreme temporality, at least as far as the psyche is concerned. It is perhaps telling that in Greek (both ancient and modern) *kairoi* (*καιροί*), the plural form of Kairos, signifies 'the times,' a less quantifiable, mostly qualitative aspect of temporality.

For Cassar, finding the opportune moment is of essence for the psyche's exploration. In his chapter 'Living *Kairos*: the time factor in waking dreams', he proposes a method as the means to such endeavour: 'Waking dreams seem to offer a unique moment, a *kairos* in the temporal process, moments in which something unique can happen or be learnt or accomplished' [p. 49]. Cassar explores the notion of time in waking dreams as described by Carl Jung in his technique of *Active Imagination*, and in Robert Desoille's *Directed Waking Dream method* (*Rêve Éveillé Dirigé*); the time factor, Cassar explains, is an important element and central characteristic in Waking Dreams since it is essential in defining these methods. Through these imaginary practices one pauses from his real life and enters an atemporal mythical realm, echoing Coleridge's most famous phrase of a 'willing suspension of disbelief', in order to be able to re-create himself. In a Waking Dream the past is re-lived in the present and the future is anticipated. Cassar also questions the future of Waking Dreams since they seem to have lost the therapeutic appeal they had in the twentieth century. He proposes a 're-awakening' of Waking Dreams in therapeutic practice, by updating them and making them more relevant to our contemporary times, through the medium of cyberspace and virtual reality; by adapting them, in other words, to the Spirit of our times.

Christopher Hauke's contribution, 'Screen time: movies, mind and the experience of time', reminds us of time outside Chronos through the use of moving images, a modern medium – a product of our times – that beautifully demolishes time's linearity. Kairos, Hauke offers, is 'the cleverness of conscious time' [p. 85], an original and highly creative perspective, akin to the medium he is exploring so skilfully. Movies present, in a unique and more pronounced way (in comparison to literary narratives), alternative temporal experiences: Time can stop, speed up, slow down, reverse, repeat. Hauke discusses memory and time as these are played out within particularly intriguing films – both in their plots and the techniques used by their makers. He analyses these movies along with new discoveries about how we perceive the moving image and what this reveals about our perception of time and its emotional effects. Using science and cutting edge research – such as Hasson's discovery of brain synchrony across the audience when watching structured films – Hauke discusses these elements of time, narrative, and moving pictures to look at the psychological implications for media entertainment and our perceptual expectations in general.

In Angela Connolly's 'Broken time: disturbances of temporality in analysis', we read that, according to Jaques (1982), Chronos and Kairos are two different temporal dimensions that cannot be experienced contemporaneously; rather, we oscillate between the two. Her contribution, a tribute to these two temporalities' vital role in the psyche, explores the psychological aspect of this oscillation. Connolly proposes two *pathologies of temporality* 'in which there is a break or rupture in the experience of time linked to a *dissociation between chronos and kairos* which in turn leads to a disturbance of the interconnection between past, present and future: "frozen time" and "dead time"' [p. 8]. In 'frozen time' chronological time dissociates, in order to avoid death or separation: the immobility of the present mimics the timelessness of the unconscious and both partners are condemned to a futile and indefinite waiting for something, anything, to happen. In 'dead time' an intense trauma destroys the experience of Kairos: time is reduced only to the meaningless progression towards death, in as much as the past catastrophe of the trauma has infiltrated the present and foreclosed the future. After presenting illuminating clinical and theoretical material, Connolly looks at the way in which Jungian analysis can restore temporality and heal the dissociation between Kairos and Chronos.

In a sense, Kairos alleviates the wounds of time. 'There is no real escape from *chronos*. But there is deed, metaphor and narrative through which experience can be transformed into significant *kairoi* that bind the time' (Dragosei 2007: 48). Analytical psychology, I would argue – and I suspect the authors of this volume would agree – is a field most suitably 'positioned' in the temporal realm of Kairos. As Ladson Hinton's chapter reminds us, the practice of psychoanalysis is in many ways a temporal process, a process of remembering, for owning and elaborating a past that gives us substance, thereby providing a basis for reflective consciousness.

Through this volume, the reader will appreciate that, even when bound by Chronos, time is not necessarily fixed by its linear parameters. Synchronicity is Jung's principle that explicitly deals with the acausal, non-linear workings of time; it can help us 'see' the workings of time and timelessness, in real-time, as it were. This concept is intimately related to our temporal experience; and that is true of Jung's own experience, as well as the unfolding of his overall theory (Yiassemides 2014/2016). Synchronicity is observed at a temporal frame within which external and eternal events coincide and meaning is attributed. The principle of synchronicity is explored in detail through two chapters of this volume: Vezzoli's as well as my chapter (Yiassemides). In both contributions, the topic is linked to clinical material as a way of exemplifying the elusive yet observable temporal nature of synchronicity.

In Caterina Vezzoli's 'Synchronicity in clinical practice', we see how the theory of synchronicity cuts across Jung's work, both in his understanding of the unconscious as well as in his vision of the transference-countertransference dialectics. Vezzoli, through astute observations and interconnections between different aspects of Jung's opus, provides a rich theoretical background for understanding the importance of synchronicity and its role in therapeutic work. The concept of emergence, she explains, is fundamental in understanding all psychological and affective phenomena that escape cause–effect explanations. For Jung, emergence and breaks in causality are extremely important concepts in understanding inherent processes in transference–countertransference dynamics, as well as implicit communications within the analytic relationship: 'In the dynamic between complexes and everyday life, aspects of the archetype may emerge in the form of symbol, and convey new meanings. We do not always have symbols; often we have pathology, symptoms, and dissociation. The energy of the archetype is responsible for synchronic phenomena and for the emergence of unexpected images, meaning and symbols that might resolve the contradictions of a situation' [p. 22]. Vezzoli presents two beautiful clinical examples that demonstrate in rich and poetic ways the therapeutic potentials of synchronicity and the ways it bridges the inter-subjective field, to transform dissociation into meaning.

In Jung's opus time is relative at large: the psyche's temporality, in all its manifestations, disregards the fixed parameters of linearity. Chronos, the Lord of clock-time, is constantly defied. Fixed and linear time gives way to relative time, where the observer is the meaning-maker, and, in essence, challenges her own temporal parameters. This non-linear temporal understanding is intimately linked to Einstein's General Relativity. 'In a sense, according to the theory of relativity, each observer is a different clock; each individual is a reality-determining location. For Einstein, the human perception of time as having a past, present and future was solely a function of consciousness, of the human mind' (Yiassemides 2014/2016: xv). Jung's temporal understanding was influenced by his contemporary scientist: Einstein's marks can be detected in Jung's time theory and particularly in his views on synchronicity's non-linear, multidimensional, and multidirectional manifestations.

In my (Angeliki Yiassemides) contribution to this volume, 'The thin curtain of Non-Space and Non-Time: synchronicity and prospective dreams; theory and clinical applications', I explore this unorthodox temporal reality as it is revealed in Jung's principle of synchronicity (an endeavour I have been working on for several years), through the example of prospective dreams, which I consider a 'subdivision' of the acausal connecting principle of synchronicity. Like Vezzoli, I use clinical material to elucidate the theoretically challenging notions behind the temporality that characterizes (and hopefully explains) instances of prospective dreaming and synchronistic occurrences within the analytic setting. By doing so I attempt to portray the unorthodox temporality that lies at the foundation of Jung's opus: Chronological time is abolished; instead 'a kind of *psychic "unit" of time* that is different – in its parameters and definition – than a unit of physical/real time (i.e. clock time)' is implied. 'This is intimately associated to the concept of relative time – as opposed to a fixed and universal "real time" – in the sense that each observer of Reality creates her own experience depending on "where" she is' [p. 64].

Culture and temporal hypostasis

Einstein's theory of General Relativity was something new and ground-breaking during Jung's epoch. Modern physics overturned Newtonian determinism and with it the assumed order of the universe. These days we often talk about time's relativity, and perhaps take the concept for granted. Nonetheless, more often than not, when we talk of 'time's relative' we are not referring to Einstein's General Relativity but rather to a more mundane (literally), non-scientific observation that has to do with its passage, its significance, and its measurement; and the fact that there is no way to universally 'gage' time.

Henri Bergson is the philosopher most notably associated with the idea that time (or *la durée* as he called it) is subject to one's conscious experience. And non-surprisingly, this was precisely the point of the epic friction between Bergson and Einstein. For Einstein time was a physical reality, something to be measured by the science of physics; his theory dealt precisely with that aspect of the nature of time. For Bergson, it was a question of our existence within a time-measuring world, or perhaps against it; time was action itself, it was related to human undertakings (Canales 2015).

There is yet another dimension of the above distinction – or perhaps 'subdivision' if we want to be more pedantic with definitions – in understanding (the nature of) time. It has to do with the way we, as a species, record history. This is beautifully demonstrated in the 2016 Pulitzer Prize winning piece (on seismology, nonetheless) by Kathryn Schulz.[3] She writes:

> On the face of it, earthquakes seem to present us with problems of space: the way we live along fault lines, in brick buildings, in homes made valuable by their proximity to the sea. But, covertly, they also present us with problems of time. The earth is 4.5 billion years old, but we are a young species, relatively speaking, with an average individual allotment of three score years and ten. The brevity of our lives breeds a kind of temporal parochialism – an ignorance of or an indifference to those planetary gears which turn more slowly than our own.
>
> (Schulz 2015)

Problems of time, Schulz reminds us, have to do with our *collective age* compared with the age of the world that hosts us. Our 'temporal parochialism,' this misapprehension, I would add, is also the source of our hubris: The assumption that we are the creators of the world upon which we rule (an assumption encouraged by Christianity's doctrine and its creation myth).

Our collective culture and its temporal substance is another theme that creatively runs through this volume. As Ladson Hinton explains in 'Temporality and the torments of time', when human beings developed a sense of future, they also gained the ability to anticipate threats from nature or their fellow beings.

They thereby created cultures that are bastions of survival, as well as places of poetry, art and religion where they could band together and reflect upon their common plight'. Cultural expressions are also mechanisms of survival, or attempts for meaning-making. And at the foundation of these collective human endeavours lies elusively, yet evidently, Time, which carries our human myth forward.

We travel through time, and in our journey we encounter mementos that once stood as contemporary artefacts – *contemporary*, in the exact meaning of the world: standing together 'with time' (con + tempus), alongside the tribe that inhabited the particular *topos* at that particular time in our species' history.

Hinton remarks about cave art: 'The cave art found in Southwest France powerfully evokes the potency of time . . . The hands seemed to reach out directly to me, their visitor of the future. My feeling of kinship with those Cro-Magnons of 30,000 or more years ago was moving and powerful. These experiences increased my wonder about time and memory as the quintessential human traits' [p. 34]. According to Hinton, in psychoanalysis we are called to remember and elaborate on the past, to give it substance and reflect upon our consciousness. He vividly amplifies this perspective through a case study which offers illustrations of a complex temporality 'that is more syncopated than linear or teleological'. Hinton's case presentation gives another perspective of the disturbed temporality of 'frozen time' encountered in Connolly's chapter: It follows the violent collapse of a man's lifelong state of 'frozen in time' where traumatic disruptions provide the possibility of a new, more open sense of temporality. Hinton's contribution also reminds us – through neuroscientific findings of memory and 'time travel' – that accepting the contradictory nature of temporal experiences can open space for increased freedom and playfulness.

Our primordial inner nature which is always within us and is evident in our cultural expressions is explored in Paul Bishop's chapter 'Just a moment? Or, the archaic as an expression of the eternal in time'. Bishop revisits the cave, first encountered in Hinton's contribution. This time however, it is the terrifying space of the cave, where memories of beasts and humans are stored. In a chapter profoundly rich with metaphors, references and depth of thought, Bishop examines the symbolisms of the cave and the forest as representing the persistence of the archaic nature within.

Jung's notion of the archaic, Bishop argues, offers a way to discuss the inner nature of humankind. Bishop, who has previously proposed a revival of Jung's notion of the archaic (Bishop 2012) seeks to extend this argumentation in close relation to the problem of time. Opening with an examination of such localized depictions of the archaic as Ovid's and Seneca's description of caves, Porphyry's analysis of Homer's cave of the nymphs, and the 1930s German film *Ewiger Wald* (Enchanted Forest), Bishop suggests that what these evocations of the archaic have in common is their emphasis on a particular *dimension* of time. This dimension is analysed with references to Plato's definition of time as an image of eternity in the Timaeus, and Friedrich Nietzsche's evocation in *Thus Spoke*

Zarathustra, of the 'midday vision' in which the Now is revealed as eternal: 'in [Jung's] account of Zarathustra's mystical experience in "At Noontide", which stands in a complex relation to Nietzsche's own account of the experience of composing *Zarathustra*, and in Jung's analysis of it, we find a gesture towards a dimension of time which can perhaps best be described as an Eternal Now. Surely this is the essence of what Jung is trying to tell us about the archaic: it precedes us, it leads us forward, it is always already there, and it can never go away' [p. 101]. As Bishop explains, for Jung, what Nietzsche had experienced, along with analogous experiences from various cultures, constituted a primordial affirmation of humankind, an instance of what Jung defined as an archetype.

J. A. Swan's 'Motion, measurement, and mechanization: the sacred-secular in temporal relics, objects, and possessions' asks poignant questions about our human temporal experience and culture: '*Who We Are* (Now), *Who We Have Been*, and developing thoughts about *Who We Are Becoming* are temporally challenging questions that illuminate the process of changing within our lifetime as individuals, and collectively, across millennia as human beings' [p. 110]. Swan's contribution is a lyrical exploration of the dialectic exchange between humans' and objects' temporal existence: It 'is a story about both cyclical and linear progression, and adaptation of physical objects to accommodate emerging culture and changing technology. It is also a story of the symbolic: Observing the patterns and rhythm of our *lives-in-time* and how we have translated our internal perceptions of temporality into the material world' [p. 109].

Architectural structures, artefacts, time-keeping devices, books and personal possessions are seen by Swan 'as temporal vessels interacting with meaning and the capacity to hold onto histories and cultures, past and present' [p. 106]. The way humans interact with these inanimate objects is a measure of time and history – personal and celestial; it is a way of 'telling time', of mapping time, of measuring and attempting to understand it. The way Swan views inanimate objects through the temporal lens is poetic yet pragmatic. Through her chapter, she treats objects as symbols of both linear and cyclical time; as vessels of our paradoxical temporality. Our interaction with objects through time becomes alive in her words. We create objects and at the same time those creations reflect our humanity and history: they have a memory and function-in-Human Time. Swan gives inanimate objects a unique and much-needed place in temporal topology: 'More so than any dying mythology or transient physical landscape in this post-Modern age it is the material objects that we inherit, acquire, collect, surround ourselves with, and live our lives within that hold our collective historical imprint and identification within time' [p. 108].

The reader must have appreciated by now the multifaceted temporal expressions explored within *Time and the Psyche*. I have organised this book in two sections, to provide a frame for these rich contributions. The first part, 'Time and the psyche: clinical practice and theoretical considerations', is composed of five chapters that illustrate the theoretical considerations through case study presentations and/or clinical considerations. Connolly, Vezzoli, Hinton, Cassar and Yiassemides approach *Time and the Psyche* via clinical work while providing extensive theoretical underpinnings. In the second part, 'Expressions of time in the psyche', temporality's multiple manifestations in the psyche and the outer world are explored by Hauke, Bishop and Swan. The lens in this section is theoretical, with rich examples from various fields and sources. Both sections explore theory as well as practical applications of the topic, each drawing on different sources of inspiration; each chapter is a unique, multi-layered exploration of Time in the Psyche, from a Jungian perspective.

Jung's theory rests on non-linear time, temporal multidirectionality, multidimensionality, and interconnection between the inner and outer realms; it operates on the assumption that beyond linearity and causality, there is an alternative temporal reality. In *Time and the Psyche*, we journey through the psyche's unorthodox temporality. And just like it is implied by the psyche's non-linear temporality, there is no end-point to this journey: 'in the end, one hardly knows how much of the old has been vaporized as dream and how much brought back to us again as dawning insight' (Campbell 1957: xi). Campbell, referring to the 1951 Eranos Meeting on 'Man and Time,' concluded: 'Our minds at this meeting were prodigiously churned' (Ibid.). It would be perhaps too ambitious to hope for the same hyperbolic sensation. Then again, it is an inevitability, while journeying through *Time and the Psyche*, to gain a deeper understanding and sense of belonging to a non-ordinary temporal reality.

Angeliki Yiassemides

Notes

1 '*πρὸς ὀλίγον ἐστὶ τὸ ζῆν, τὸ τέλος ὁ χρόνος ἀπαιτεῖ.*'
The epitaph of Seikilos is the oldest surviving complete musical composition, including musical notation, worldwide, the most likely date being the first century AD.
2 Pausanias (1918) claims Kairos was the youngest divine son of Zeus, whereas Battistini (2005) suggests he was the son of Oceanus.
3 Kathryn Schulz was the 2016 Pulitzer Prize winner for the category Feature Writing.

References

Augustine, S. (1998). *Confessions*, trans. H. Chadwick, New York, Oxford University Press

Battistini, M. (2005). *Symbols and Allegories in Art*, trans. S. Sartarelli, California, Getty Publications

Bishop, P. (ed.) (2012). *The Archaic: The Past in the Present*, London and New York, Routledge

Borges, J. L. (1974). Nueva refutacion del tiempo, in *Obras Completas de Jorge Luis Borges*, Buenos Aires, Emece Editores

Campbell, J. (1957). Editor's foreword, in *Man and Time, Papers from the Eranos Yearbooks*, ed. J. Campbell, trans. R. Manheim, Bollingen Series XXX, New York, Pantheon Books

Canales, J. (2015). *The Physicist and the Philosopher: Einstein, Bergson, and the Debate that Changed our Understanding of Time*, Princeton, New Jersey, Princeton University Press
Dragosei, A. (2007). Frozen time and endless analyses: a response to David Tresan's 'Thinking individuation forward'. *Journal of Analytical Psychology* 52: 41–45
Jaques, E. (1982). *The Form of Time*, New York, Crane, Russak
Kerényi, C. (2006). *The Gods of the Greeks*, London, Thames and Hudson
Pausanias (1918). *Description of Greece* (Books 1–2), trans. W. H. S. Jones, Cambridge, Massachusetts, Harvard University Press
Rose, J. (1997). Distortions of time in the transference: some clinical and theoretical implications. *International Journal of Psychoanalysis* 78: 453–68
Schulz, K. (2015). The really big one: An earthquake will destroy a sizable portion of the coastal Northwest. *The New Yorker* (July 20, 2015), www.newyorker.com/magazine/2015/07/20/the-really-big-one
Yiassemides, A. (2014/2016). *Time and Timelessness: Temporality in the Theory of Carl Jung*, London and New York, Routledge, Taylor and Francis Group

ACKNOWLEDGEMENTS

This volume is the product of the desire to collect the thoughts and experiences of scholars and practitioners in the field of Analytical Psychology on the topic of temporality. It began as an international conference, on the island of Cyprus on 27 September, 2014. I am grateful to all who contributed in making this possible.

My heartfelt gratitude to Maria Vilanidou who supported my idea and helped in successfully and beautifully co-organizing the conference *Time and the Psyche: Jungian Perspectives*. To all the speakers, who travelled to Nicosia, Cyprus to present their work, I am eternally grateful: Angela Connolly, Caterina Vezzoli, Christopher Hauke and Laner Cassar, you made the conference and this volume possible with your generosity. A warm thank you to *Centro Italiano di Psicologia Analitica*, *Instituto per l'Italia Meridionale e la Sicilia* and the *Malta Depth Psychological Association* that supported *Time and the Psyche*, the conference, with their member-speakers. Three more contributors were added to complete this volume: Ladson Hinton, Paul Bishop and J. A. Swan, a warm thank you for your time and willingness to contribute so that *Time and the Psyche* is richer and more comprehensive. I am grateful to all the authors for your willingness to participate in this edited volume, and for the production of such original and enriching contributions.

Grateful acknowledgement is made to *the Journal of Analytical Psychology* for permission to reprint the following previously published material: Hinton, L. (2015). Temporality and the torments of time. *Journal of Analytical Psychology* 60(3): 353–70.

PART A

Time and the psyche: clinical practice and theoretical considerations

> I am a time being. Do you know what a time being is? Well, if you give me a moment, I will tell you. A time being is someone who lives in time, and that means you, and me, and every one of us who is, or was, or ever will be.
>
> (Ozeki 2013: 3)[1]

1 Ozeki, R. (2013). *A Tale for the Time Being*. London, Canongate Books.

1

BROKEN TIME

Disturbances of temporality in analysis

Angela Connolly

Introduction

There seems to be something about time and temporality that defies definition as we can begin to see from the famous phrase of Saint Augustine in *The Confessions*, 'If nobody asks me, I know: but if I were desirous to explain it to one who should ask me, plainly I know not' (Augustine 1913: 239). Even today, although over 15 centuries have passed, physicists, philosophers and depth psychologists are still struggling with this problem of what we mean when we speak of time and temporality. Indeed, there is something uncanny about time as we can see from the fact that, as Marie Bonaparte notes, 'the ideal of many philosophical schools is the destruction and suppression of time' (Bonaparte 1940: 453). This uncertainty about the true nature of time can also be linked to the fact that over the centuries humans have always tended to think of time in terms of oppositions between two different experiences of temporality, and, in particular, the opposition between a subjective mental experience of time and an objective experience of time as something external to the human subject.

The Greeks distinguished between two different temporalities: *chronos* and *kairos*. *Kairos*, according to Elliot Jaques, in his 1982 book *The Form of Time*, denotes the lived experience of time, the time of human intentionality, purpose and goals in which past memory, present perception and future desire flow together; whereas *chronos* refers to the measurable time of succession, the conscious perception of the passage of units of time with its asymmetry of past and future and the idea of the irreversibility of 'the arrow of time'. For Jaques, these two different dimensions of time cannot be experienced contemporaneously but rather there is a cognitive oscillation between the two, organizing our behaviour in relation to time and our experience of temporality. James Rose too distinguishes between *chronos*, clock time and *kairos*, 'the sense of a special time that is significant and meaningful' (Rose 1997: 453), but he suggests that this distinction 'should not be seen as a dichotomous split but as two elements struggling for a dynamic equilibrium within a dialectic in which each assumes and implies the other' (Rose 1997: 464–65).

Modern physics: the illusion of time

Until the advent of modernity, humanity always recognized the reality of the passage of time and saw time as an absolute value; with the advent of modern physics even this very reality is, in a certain sense, negated. As Einstein wrote in a letter of condolence to the sister and son of Michele Besso following his death, 'For us who believe physicists, the distinction between past present and future is only a stubbornly persistent illusion' (Einstein and Besso 1979). Rather than seeing space and time as absolute, as Newton did, Einstein made use of Minkowski's concept of a four-dimensional space–time continuum with time as the fourth dimension to stress the relativity of time and space. In a similar fashion, Roger Penrose notes that the idea of the passage of time:

> is something so familiar to us that it comes as a shock to learn that our wonderfully precise theories of the behaviour of the physical world have had, up to this point, virtually nothing to say about it. Worse than this, what our best physical theories *do* say is almost in flat contradiction with what our perceptions seem to tell us about time.
>
> (Penrose 1995: 384)

The physicists Hans Atmanspacher and Hans Primas, who approach this problem from the view point of dual aspect monism and who are very open to the views of Jung, have a more nuanced approach to time which also considers its subjective experience. They distinguish between the time of a mental domain and the time of a physical domain, both of which are hypothesized to come into being through the breaking of the symmetry of an original ontic reality. As they write:

> Time appears to have two aspects which one may call physical and mental. They refer to two mutually exclusive *complementary* ways of looking at a fundamental ordering principle . . . at the psychophysical neutral level of the unus mundus there is no time at all. Time emerges as an epistemic ordering parameter due to a symmetry breaking of the primordial reality referred to as the unus mundus. The associated distinctions leads to a tensed and tenseless domain.
>
> (Atmanspacher and Primas 2006: 25)

In this vision, 'tensed' refers to mental time where time is organized into past, present and future and is intuitively imagined in terms of flux or of a unidirectional movement from past to future as in the metaphor of the arrow of time, whereas physical tenseless time is limited to the relations of 'earlier than', 'later than', and 'simultaneous with.' This physical time, as the authors point out, 'is not the time we experience mentally but a homogeneous *parameter time* referring to an *external clock* carried by an external observer which is not part of the physical system under discussion. Moreover there is nothing that *flows* or *passes* in this physical time' (Atmanspacher and Primas 2006: 23). For Atmanspacher and Primas the holistic nature of the *unus mundus* implies that the states describing the material and the mental domains are entangled; from this Hans Primas (2003) argues that 'the tensed time of the nonmaterial domain is synchronized with the tenseless time of fundamental physics by quantum correlations' (Atmanspacher and Primas 2006: 114). In other words, mind and matter may be related via a temporal domain, serving as an interface between atemporal material and temporal mental domains.

Phenomenology: the time of consciousness

Even though from the point of view of modern physics time is an illusion, temporality – i.e. the subjective experience of time – and what Husserl terms 'internal time-consciousness' are essential to the experience of consciousness itself; for if each moment of consciousness were not temporally connected with previous moments of consciousness, there would be no continuity of memory, judgement and identity. As the phenomenologist, Shaun Gallagher, puts it

> There is, implicit in the very nature of consciousness (no matter if it is perception, memory, imagination, a train of conceptual thought etc.) a binding of one moment to the next. This binding process is what Husserl calls 'retention' in regard to past moments of consciousness, and 'protention' in regard to the future. Husserl's model explains not only how the perception of a temporal object such as a melody, is possible, given a changing stream of consciousness, it also explains how consciousness unifies *itself* across time.
>
> (Gallagher 2007: 609)

Psychoanalysis: unconscious time

Like many philosophical schools that operate under the assumption of 'the destruction and suppression of time' (Bonaparte 1940: 453), psychoanalysts have tended, until relatively recently, to side-line the significance of time and temporality in therapy preferring to give Freud the last word on the subject. As Borgin notes, 'As time is the very medium of psychoanalytical therapy (the time of the session, the time of personal history) it is unsurprising that the category has largely escaped objective consideration in psychoanalytical theory' (Borgin 1996: 215).

For Freud, temporality is an essential characteristic of consciousness: he contrasted the conscious experiences of time with the timelessness of the unconscious, stating that 'The processes of the system Ucs. are *timeless*; i.e. they are not ordered temporally, are not altered by the passage of time; they have no reference to time at all' (Freud 1915: 187). Again in 1932 he wrote:

> There is nothing in the id that could be compared with negation; and we perceive with surprise an exception to the philosophical theorem that space and time are necessary forms of our mental acts. There is nothing in the id that corresponds to the idea of time; there is no recognition of the passage of time and – a thing that is most remarkable and awaits consideration in philosophical thought – no alteration in its mental process is produced by the passage of time. Wishful impulses which have never passed beyond the id, but impressions, too, which have been sunk into the id by repression, are virtually immortal; after the passage of decades they behave as though they had just occurred. They can only be recognized as belonging to the past, can only lose their importance and be deprived of their cathexis of energy, when they have been made conscious by the work of analysis, and it is on this that the therapeutic effect of analyst treatment rests to no small extent.
>
> (Freud 1937: 74)

From Bonaparte onwards, however, analysts have begun to take issue with Freud's radical claim, suggesting that the unconscious is not timelessness but rather it is characterized by *a different kind of temporality*. As Charles Hanly states:

> In themselves unconscious processes are no less temporal than conscious psychic processes. It is just that they function differently and one of the differences is that they are not, once established, easily subject to the influences of development and relations. But they are not at all immune to change in the sense in which, for example, Kantian *noumena* are. Moreover, they can be changed either by good fortune in life of an appropriate kind (the inheritance of wealth will not do, luck in love might) or by psychoanalysis. What cannot be changed by time, what has not been changed by life, still can be changed by psychoanalysis as Freud (1923) recognized. Trauma can cease causing psychopathology. Memories can cease to act as though they were current rather than past experiences. When they do, they take their place in the temporal sequence of the individual's life experience that they always had.
>
> (Hanly 2007: 8)

Matte Blanco has put forward a rather sophisticated vision of the role of time in the unconscious. For Matte Blanco, whereas the logics of ordinary thinking is characterized by asymmetrical relations, unconscious thought processes are characterized by symmetrical logic which is incompatible with the concepts of time, space and movement: 'Time involves awareness of an ordered sequence. With symmetrisation, order which is by its essence asymmetrical, disappears, and so will awareness of time' (Rayner and Tuckett 1995: 26). In this conceptualization, the mental domain is structured by an infinite number of strata, which go from the most conscious level – characterized by asymmetrical processes through levels which become increasingly symmetrised – to the most unconscious level, the 'indivisible mode', where the degree of symmetrisation is such that any awareness of time and space become impossible.

Jean Laplanche distinguishes between four different levels that need to be taken into account in any consideration of time: cosmological time, historical time, biological time of the living being, and time of the human subject which refers 'to the capacity which the human being has of creating, of secreting [her or his] own time' (Laplanche 1992: 162).

Jung: the relativization of time

There are important similarities between Matte Blanco's model of the psyche and that of Jung in that both imagine levels of the mental domain that go from the conscious level to that of the most unconscious level. But whereas Matte Blanco organizes these levels in terms of thought processes, Jung sees them in terms of archetypes and of a gradual relativization of time and space until one arrives at the psychoid level of the indivisible *unus mundus*, where time and space no longer exist. Jung's approach to time and temporality is mainly developed though his ideas on synchronicity in as much as synchronicity depends on the relativization of time and space which leads to an entanglement of the material and the mental domains in a way that correlations become possible. As both Main (2004) and Yiassemides (2014/2016) point out however, Jung is often inconsistent and ambiguous in his references to time and to simultaneity in relation to synchronicity, something which Wolfgang Pauli pointed out to Jung himself. If we go back to Jung's definition of synchronicity as 'the simultaneous occurrence of a certain psychic state with one or more external events which appear as meaningful parallels to the momentary subjective state' (Jung 1952: para. 850) the first problem is that Jung is not always clear about what kind of time he is referring to when he talks about *simultaneity*.

Yiassemides (2011) points out that Jung is in fact referring to a very particular type of *chronos*. In a letter to Pauli, Jung writes:

> It looks as if the collective character of the archetypes would manifest itself also in meaningful coincidences, i.e. as if the archetype (or the collective

> unconscious) were not only inside the individual but also outside, viz. in one's environment, as if the sender and percipient were in the same psychic space, or in the same time.
>
> (Jung 1974: 46)

As Yiassemides notes, 'here Jung seems to imply a temporal (as well as spatial) parameter that behaves *as if in conscious time*, but is nonetheless different from it' (Yiassemides 2011: 464–65).

In other words, the experience of time from the point of view of the psyche is very different from that of consciousness and, as Yiassemides points out, it is much closer to the Chinese conceptualization of time which is seen more as a field of time rather than as 'an arrow of time, where cause precedes effect' (Yiassemides 2011: 465). As Gieser says, in Chinese tradition, 'Time is above all a quality, expressing a pattern which has a significance or meaning' (Gieser 2005: 279).

As psychoanalysts have begun to distinguish between different temporalities, attention has been increasingly focused on how different psychopathologies can lead to dissociations between objective and subjective time, and the way in which these different disorders can subtly alter the transference and countertransference dynamics and the experience patients have of the particular temporality of analysis. What I propose in this work is to look at two distinct *pathologies of temporality* in which there is a break or rupture in the experience of time linked to a *dissociation between chronos and kairos*, which in turn leads to a disturbance of the interconnection between past, present, and future: 'frozen time' and 'dead time'. To this end, I will offer some considerations about how the development of the sense of temporality unfolds and the role of metaphor in our conceptualizations of time.

The development of temporality

The way in which we develop both the subjective sense of temporality and the capacity for abstract conceptualization of time is intimately linked to our embodied experience of the movement of objects and of our own movements through space. It is exactly this embodied experience of movements in space that provides the spatial metaphors that allow us to experience and conceptualize time; as Julian Jaynes states, 'You cannot, absolutely cannot, think of time except by spatializing it. Consciousness is always a spatialisation in which the diachronic is turned into the synchronic, in which what has happened in time is excerpted and seen in side-by-sideness' (Jaynes 1996: 60). Sheets-Johnstone has suggested that before infants can begin to develop the sense of *befores*, *nows* and *afters*, they are aware of the qualitative flow of events that make up the contours of lived experience (Sheets-Johnstone 1999: 157). This qualitative aspect of the different rhythmic movement of objects and of the body through space are essential in attributing qualitative aspects to our experience of time. Lakoff and Johnson (1999) have analysed in detail in *Philosophy in the Flesh* the two principal spatial metaphors through which infants begin to develop a subjective sense of temporality which eventually form the basis for the abstract conceptualization of time: *the moving time metaphor* and *the moving observer metaphor.* According to Johnson in *The Meaning of the Body*, 'The first spatialization understands discrete times metaphorically as objects moving towards a stationary observer, first in front of the observer, the passing her, and finally moving further and further away behind her' (Johnson 2007: 29). In this way, the movement of objects allows the child to begin to construct a metaphorical understanding of the passage of time. In the second spatialization, the experience of an observer moving through space again facilitates the metaphorical understanding of the duration of time: the distance moved becomes mapped onto the amount of time passed.

These two different spatial metaphors are fundamental in the creation of our different experiences of time. In the first, the observer is static and it is time that moves, whereas in the second it is the observer who moves through time. I would suggest that the sense of time as something external which moves past us contributes to the experience of *chronos*, while the sense of time as something we move through is fundamental in the creation of *kairos*. If all goes well, children will be able to gradually bring together these two different metaphors of time to create that dialectical relationship between *chronos* and *kairos*; between the experience of external, objective time that marks out the span of our lives as living organisms, and the subjective experiences, conscious and unconscious, of moving through time and of being able to imaginatively create our own time.

This dialectical relationship is essential if we are to be able to bring together past, present and future, to weave them into a continuum which for Heidegger is the hallmark of authentic temporality. If, however, because of developmental failures and deficits or because of traumatic life events, the subject is unable to bring together *chronos* and *kairos* or if they have become dissociated, then the continuum between past, present and future is broken and pathologies of temporality will manifest. In this chapter I will be looking at two psychopathologies of temporality: 'frozen time' and 'dead time.'

Frozen time

In patients where time becomes frozen, the dialectical relationship between *chronos* and *kairos* breaks down or has never been established; the result, as James Rose points out, is the refusal or denial of *chronos* and 'the creation of a space-time of *kairos* as a defence against the insistence of *chronos*' (Rose 1997: 453). The result is not so much timelessness as the feeling that chronological and developmental time has been frozen.

A patient of mine who has been in analysis for fifteen years because of a severe masochistic perversion, had the following dream after four years of analysis:

> I am visiting my mother's homeland and I meet my aunt but I can't find my little cousin and my aunt says that she doesn't know what has happened to him. I find out that he has gone to the house of my mother's parents but that he had got cold and had turned back. I look everywhere for him and I find him in a frozen alarm clock. I wait until the ice melts to see if my cousin comes out.
>
> (Connolly 2013: 648)

This sense of frozen time is conveyed graphically in Faulkner's *The Sound and the Fury* in which, as Sartre notes, time is distorted because it is decapitated, 'deprived of its future, that is of its dimension of deeds and freedom' (Sartre 1955: 84). For Faulkner, according to Sartre, 'time is, above all, that which separates' (Ibid.) and the only solution for certain individuals is to freeze time, to stop the clock in order to foreclose the possibility of separation. The result is a present that is completely subsumed by the past and where the possibilities of the future are forever denied: 'In *The Sound and the Fury*, everything has already happened' (Sartre 1955: 81).

When the sense of time is distorted, however, there is a risk that analytical temporality can break down and that a very particular kind of transference/ countertransference dynamic can take over. According to Gilda De Simone (1997), temporality is crucial to the very possibility of separation and of ending analysis and when there is a dissociation or a denial of chronological time, fantasies of interminability or of breaking off become a central feature of the analytical relationship. If the countertransference is not adequately reflected on because of the analyst's own separation anxieties and refusal of incompleteness and death, this can lead her to collude with this denial of time.

In these cases, the temptation on the part of the analyst to react against the immobility and the denial of time by artificially introducing temporality into the analysis can be very strong, but as De Simone points out, this is a unilateral, inappropriate intention on the part of the analyst; the result is usually a kind of perversion of the transference on the part of the patient, and to something resembling a perversion of the countertransference on the part of the analyst (De Simone 1997: 37). This kind of analytical acting out leads only to the imitation of a good analysis in which the possibility of transformation is absent (as we can see clearly in the case of the Wolfman, for example). In the end, I believe the decision to terminate can only belong to the analysand and the ethical responsibility of the analyst resides rather in her capacity to become aware of, and to reflect back to the analysand, movements that indicate that he is ready to terminate.

Elvio Fachinelli, one of the most creative and original Italian psychoanalysts, in his book *Claustrophilia* describes what he calls the 'claustrophiliac situation', a situation in which 'the immobility of the present mimics the atemporality of the unconscious' (Fachinelli 1983: 190) and analyst and analysand become frozen into an indefinite and futile waiting for something, anything, to happen; the failure of the process becomes masked by an idealization of analysis and by an inflated overvaluation of its possibilities. In the last analysis, 'the creativity of the analytical process and its ability to bring about transformation, depends on the capacity of the analyst to maintain, at least within himself, the opposition between *chronos* and *kairos*, between terminability and interminability' (Connolly 2007: 43–44).

Dead time

In dead time – a term I use to distinguish from the experience of 'the death of time' to which Yiassemides (2014/2016) refers to – a traumatic event which can be more or less intense, leads to a breakdown in representation and in metaphorical capacity which in turn leads to the destruction of the experience of *kairos*. This results in time becoming reduced to a purely concrete chronological experience.

Ilse Grubich-Simitis, in her ground-breaking article on her work with children of Holocaust survivors, *From concretism to metaphor*, notes that in the psychotic universe of the extermination camps, 'the occurrence of totally senseless events – events that have no explanation, indeed cannot be explained . . . undermines the metaphorical and the non-metaphorical use of speech as well as the structuring of time in past, present and future' (Grubich-Simitis 1984: 307).

Without the experience of *kairos*, time becomes literalized, reduced only to the meaningless progression of a chronological time, deprived of all its symbolical power. When time is reduced only to chronological progression, the way in which the subject experiences death also becomes radically modified. In *The Soul and Death*, Jung (1934) stresses the symbolic importance of the acceptance of death and of the stopping of the biological clock. As Yiassemides states, Jung's 'adoption of a symbolic view allowed for his theory's expansion beyond anxiety of physical death. On the symbolic plane, death is a psychic experience pregnant with hope and vitality' (Yiassemides 2014/2016: 146).

The breakdown of symbolic capacity, which is found in intense trauma and in intergenerational trauma, leads to an incapacity to think of death as a symbolic experience. If time is reduced to *chronos*, and death is experienced only as a concrete event to which the individual can only passively submit, then all too often (as we know from the experiences of survivors such as Jean Amery and Primo Levi) the only way to insert intentionality into time and to recover the possibility of the symbolical meaning of time is to destroy *chronos* through suicide. This is summed up in a phrase from Faulkner's *The Sound and the Fury*, 'Time is dead as long as it is being clicked off by little wheels; only when the clock stops does time come to life' (Faulkner 1931/1964: 81).

The phrase 'the death of time' was first used by Elie Wiesel (1975: 314–15) to describe the temporal experiences of the inmates of the death camps. Kijak and Funtowicz have described the profound dissociation in temporality that occurred in survivors of extreme situations leading to 'the simultaneous coexistence of two aspects of the ego: one part of this ego continues "living" in the death camp stripped of all its defenses; the other part, "adapted" to the new reality, behaves . . . was if it were able to love, to hate, to struggle, to work, making projects or becoming ill' (Kijak and Funtowicz 1982: 30). As the Spanish writer Jorge Semprun, a survivor of Buchenwald, writes, it is as if 'I never left, in spite of all appearances, and which I would never leave, despite the masquerades and make-believe or life' (Semprun 1997: 153).

The same kind of disturbance of temporality can be found in the children of survivors of intense trauma, as the work of Judith Kestenberg has shown: she writes of a 'time tunnel' (Kestenberg 1982: 83–102). In a similar tone, Haydee Faimberg speaks of 'a telescoping of generations,' a 'tyrannical intrusion of history' (Faimberg 1988: 99). Often the survivors identified children born after the Holocaust with the deceased and the children themselves identified with the dead. In this way, they became a kind of 'revenant.' In such cases, there is a "vampiric identification" in which the child of survivors become imprisoned in the parents' trauma, neither dead nor alive, unborn, in an imageless, timeless condition, condemned to repeat what they themselves have not experienced' (Wilgowicz 1999: 1063).

A similar phenomenon was again found in interviews carried out with second and third generation Jews whose grandparents or parents were inmates of death camps. Raffaella Di Castro, a philosopher, in *Testimonies of the Non-Experienced*, collected and analyzed the stories of 23 third generation Roman Jews, whose relatives had been victims of the Fascist persecutions. She speaks of 'a traumatic *topos* in the construction of identity, a story which has the power of a "myth" and which risks being mistaken for one, becoming merely "pre-history" or "post-history" of the mind, passed on from generation to generation and of the need to make sense of these paradoxical and difficult memories of an experience that has not been lived but transmitted' (Di Castro 2010: 326–27). As I remarked in a previous paper, 'Thus the death of time creates a dissociation between the chronological time of history and the subjective time of memory with the result of the creation of a history without memory, history as abstract dead facts, and a memory without history, purely subjective, mythical and therefore ineffective for the creation of meaningful narratives' (Connolly 2011: 612).

In the same way, the capacity to perform dream work, in order to represent the trauma and render it thinkable, also fails. Trauma dreams, as Jung notes in the *Kindertraume*, are pure repetitions of real events with no dream work:

> The dream is never a mere repetition of previous experiences, with only one specific exception: shock or shell-shock dreams, which sometimes are completely identical repetitions of reality. That, in fact, is a proof of the traumatic effect. The shock can no longer be *psychified*. This can be seen especially clearly in healing processes in which the psyche tries to translate the shock into a psychic anxiety situation.
>
> (Jung 2008: 21)

In trauma dreams, the coercive and repetitive nature of the non-memories blocks the capacity to perform dream work and to transform the images, for here the dream images are not metaphoric representations but iconic presentations of the transmitted trauma. This failure of representation leads, as Botellas writes, to 'a disorganizing state at the limits of the psychic and inaccessible to elaboration which may be described as *non-representation*' (Botella and Botella 2005: 125). If the reconstruction of the experience of *kairos*, of the subjective experience of temporality is not to finish in suicide then the only possibility that remains is the re-metaphorization of time through what Laub and Podell (1995) call the 'art of trauma'. This can only come about through artistic creation or through analytical work. As Marcelo Vinar states, in his discussion about the possibility of overcoming trauma, 'only the skill of a poet or a patient in transference can go some way to achieving this by using metaphor as best he can to express the emotional intensity and the incandescence of the experience where words so often fail' (Vinar 2005: 315).

Restoring time

In analysis, the analyst has three fundamental tools at her disposition to help her restore the interplay between *kairos* and *chronos*: dream work, the work on the transference–countertransference dynamics, and the very particular temporality of analysis. The failure of representation and the incapacity to perform dream work in traumatized patients requires the analyst to dream the patients' dreams. Botellas refer to this as 'the analyst's work of figurability, a work that can only come about when the analyst is able to accept a regression to the most unconscious levels of the psyche in which he or she becomes the analysand's double' (Botellas 2005: 71).

As I have already discussed the first two tools in previous papers (Connolly 2011, 2013), I will concentrate here on the *temporality of analysis*. A fundamental part of the possibility of performing analytical work is the temporal character of the analytical setting. As Sabbadini notes:

> [T]he timeless quality of the content of analysis is determined by, and in constant interaction with, such formal time arrangements, set by the analyst and altered only under exceptional circumstances. It is this contrast of temporalities that shapes the analytical encounter, modulating its rhythm and punctuating its discourse. Each of these temporalities is unthinkable without the other.
>
> (Sabbadini 1989: 307)

To illustrate how particular temporalities of analysis work I will give a brief account of an analytical session where the experience of time was the central feature.

Chronos' child

Jim, a young man in his late twenties, came in to analysis because of panic attacks which were becoming increasingly incapacitating. Jim's life had been relatively uneventful until the year before when the car he was driving became involved in an accident through no fault of his own. He had a successful career as a website designer, was married with one child, and had never suffered from anxiety until the accident although, as it gradually emerged, he had always been slightly obsessive. After this accident, however, whenever he found himself in a situation where he felt blocked and unable to move, he would begin to feel intense panic until he was able to extricate himself from the situation. The attacks had begun when he found himself blocked in traffic jams, but soon extended to situations such as theatres or restaurants when he would feel hemmed in by other people; the idea of taking a train or an aeroplane had become intolerable although previously he had loved travelling to exotic places. The other precipitating event which had happened shortly before the accident had been the successful cancer operation undergone by his father.

As the analysis proceeded, it gradually became clear that Jim's experience of time had been radically changed by the accident and the possibility of his father's death. As soon as he felt immobilized, Jim had the distinct sensation that time had begun to slow down, or even that it was no longer moving, and he would obsessively control his watch to see how much time had passed. At an unconscious level, Jim equated the idea of time stopping with death: he experienced time only in terms of a chronological time that was relentlessly moving him towards the moment when it would finally stop. Jim had therefore lost all sense of subjective time, the time we move through actively, the time we create through our actions and imagination, and could only defend himself against the dangers of chronological time by controlling it, by dividing it into separate, unconnected instants by controlling his watch. When phobic patients 'feel trapped or persecuted by time' (Boschan 1990: 342), the solution – according to Fachinelli – is 'the segmentation of concrete time, of time as flux and the individual form of action, into a series of instants which tend to become ever smaller' (Fachinelli 1992: 23).

I will use the transcription of parts of a session with Jim, which took place a year and a half later, to illustrate how the different temporalities and spatial characteristics of the analytical setting have modified Jim's experience of time. I have deliberately used the present tense in this description, as I feel it better conveys the flow of the session.

Jim arrives at the session which came after a session he had missed for reasons of work; unusually, he is a few minutes early. As is his habit, he begins to speak at once, saying he has been fine despite the missed session; nothing has gone wrong. Indeed, he feels that his life is a little *flat* at the moment as nothing much has been happening. That morning coming here there was no traffic and everything had moved fluidly. In fact, he has the sensation that sometimes he is looking for something to happen, that he would almost like it if something were to go wrong. The only symptom left is that he still has problems with his breathing and sometimes has the sensation that there is something in his throat that goes neither up nor down. I comment that it is sometimes reassuring to control reality, but that this blocks the possibility of being surprised. Jim continues talking about how routine his day is, but that now he hardly ever looks at his watch, only when he feels hungry or in similar situations. He had been in the country that weekend and he had taken his watch off altogether. He felt really good without it and he hadn't even had problems with his breathing. I comment that it sounds as though he feels that he doesn't need to control time anymore, and that he seems to be more aware of the time governed by his biological rhythms such as being hungry or sleeping; people, I offer, invented watches when they began to need to control time. Jim agrees and says that now, even if he wakes up during the night, he no longer needs to look at his watch. Before, he used to look at it and think about how long he still had to sleep and worry about whether he would get back to sleep.

He continues talking about sleeping, saying that it annoys him when his wife sometimes rocks their four-year-old son to sleep. He has tried to convince her that she is spoiling him; even though she continues to do so, at least now she knows it's not right. I comment that small children very often have difficulty sleeping and that there is something comforting about rocking. Jim says that he remembers that when he was a child he always went to sleep by himself without any problems. He could hear the TV in the next room and that was enough; he used to fall asleep with the TV on but recently he tends to switch off the TV before he goes to sleep. He was never frightened of the dark when he was a child, and he still never needs a light on when he gets up at night. When he was driving in the country that weekend however, there were no lights and he felt disorientated, as though he wasn't in control. He started talking to the others in the car and then he was able to drive without thinking.

At that point, Jim begins to look around the room and notices that I have changed the order of the books. I agree that I have tidied them up a bit. He remarks that he always notices the colours of my books, and remembers that a green one had been in another place. He says that he particularly likes the lilac colour of another book, and that he had noticed that it was in Russian. He always notices the distribution of a room and can reproduce it almost perfectly; he usually forgets only small details. The other day he went to a newsagent where he hadn't been for some time and he noticed at once that he had changed things. He commented on it and the newsagent said that even most of his habitual clients hadn't noticed the change.

I ask how he feels about the change in this room and he comments that it doesn't bother him: even if some things change the room is still the same. There's something about this room that is different from other rooms. It seems to him as if I am looking at him through a kind of a cone, with the narrow end pointing towards me and the wide end towards him. If he looks at me the distance increases a bit, like in a telescope when you look from the opposite end. It's not an unpleasant feeling, just different from reality. I comment that distances change in analysis, depending on the moment and the felt movement between presence and absence. Jim replies that he has noticed that other people don't understand the fact that analysis is different. He likes this room, he has always felt comfortable here but he wouldn't have liked it if he had to use the couch and be able to see me. He doesn't think he would have stayed if he had to use the couch; when I had told him he had a choice he felt relieved.

Winnicott underlined the importance of the mother's capacity to provide continuity and not to hurry the child's development to enable him to 'catch hold of time, in to get the feeling of an internal personal going along' (Winnicott 1958: 161). If this does not happen, then 'time will be an alienated time, belonging to the realm of the false self; instead of us possessing time, time possesses us' (Boschan 1990: 340). In this session, we can see how a probable lack of synchronization between the mother's and Jim's rhythms – between Jim's dependency needs and the mother's need for a child that was autonomous and capable of early self-control – already created in Jim the feeling of not being able to move freely through time, of not being able to insert his actions and intentionality into time. This initial deficiency in the development of the sense of subjective time opened the way to the subsequent loss of the sense of *kairos* following the later trauma and led to the use of obsessive mechanisms to control the movement of chronological time. By using an analytical setting that ensured temporal and spatial continuity that provided Jim with a secure holding environment, adapted to his need to feel in control over the analytical time and space, we can see how he has been able to gradually relinquish his obsessive observation of chronological time.

In this session, Jim's emphasis on the stability of the analytical space, despite the changes he observed, suggests that it was above all the continuity in the spatial characteristics of the setting that has allowed him to begin to develop a more subjective sense of time. This is emphasized by the use Jim made of the spatial metaphor of looking through the opposite end of the telescope to communicate his experience of distance – of the rhythms of presence and absence of another – that are adapted to his needs. It would seem that it is exactly this kind of experience that allowed him to get a hold of time, to begin to develop the feeling of internal personal 'going along'. This suggests that in patients such as Jim, who have lost the sense of subjective time, we need to look carefully at how the continuity of the particular spatial characteristics of the setting can open the way to restoring the sense of *kairos* and the dynamic equilibrium between *chronos* and *kairos*.

Conclusion

Man's relationship with time and temporality is both complex and ambiguous. We intuitively feel that we know perfectly well what time is; at the same time, we have always struggled to find definitions that can allow us to encompass the many different aspects of our relationship to time: on the one hand, the chronological time of history and of life span – *chronos* – and on the other, the subjective, imaginal time of *kairos*. The sense of time is not automatically given, but it develops through a process of metaphorization, through a series of spatial metaphors linked to the movements of objects in space and to our own bodily movements and actions. Gradually, these two metaphoric representations of time come together in a dialectical relationship and time becomes structured in past, present and future. This developmental process can fail to develop or can be disturbed; the dialectical relationship breaks down, and pathologies of temporality develop.

In *frozen time* the inability to accept separation and death leads to a refusal of *chronos*; that the subject becomes blocked in a purely subjective experience of time which mimics the timelessness of the unconscious. All too often, the same refusal of chronological time can lead in analysis to situations of interminability. Only the analyst's capacity to hold in mind the dialectical relationship between *chronos* and *kairos* can unblock the situation.

Severe trauma, such as that experienced in the death camps, leads to the 'death of time' (equally relational trauma of the kind experienced by Jim), and can also lead to the loss of the subjective sense of time of *kairos*: all that remains is the concrete experience of chronological time. In extreme trauma, the emphasis must inevitably be placed on the analyst's ability to accept the emotional impact of the concrete reality of the trauma, and to represent it in her own mind to open the way for a gradual process of re-metaphorisation and for the restoration of the subjective experience of *kairos*. In less extreme traumas, however, it is above all the *spatial and the temporal characteristics of the analytical setting* that are the fundamental factors in healing the dissociation between *chronos* and *kairos*.

References

Atmanspacher, H. and Primas, H. (2006). Pauli's ideas on mind and matter in the context of contemporary science. *Journal of Consciousness Studies* 13: 5–50

Augustine, S. (1913). *Confessions, Book 11*, London, Heinemann

Bonaparte, M. (1940). Time and the unconscious. *International Journal of Psychoanalysis* 21: 427–68

Borgin, V. (1996). *In Different Spaces: Place and Memory in Visual Culture*, Berkeley, California, University of California Press

Boschan, P. J. (1990). Temporality and Narcissism. *International Journal of Psychoanalysis* 17: 337–49

Botella, C. and Botella, S. (2005). *The Work of Psychic Figurability: Mental States without Representation*, Hove, UK, and New York, Brunner-Routledge

Connolly, A. M. (2007). Frozen time and endless analysis: a response to David Tresan. *Journal of Analytical Psychology* 56(5): 607–27

——. (2011). Healing the wounds of our fathers: intergenerational trauma, memory, symbolization and narrative. *Journal of Analytical Psychology* 56(5): 607–27

——. (2013). Out of the body: embodiment and its vicissitudes. *Journal of Analytical Psychology* 58(5): 636–57

De Simone, G. (1997). *Ending Analysis: Theory and Technique*, London, Karnac Books

Di Castro, R. (2008). *Testimoni Del Non-Provato: Ricordare, Pensare, Immaginare La Shoah Nella Terza Generazione*, Roma, Carocci Editore

Einstein, A. and Besso, M. (1979). *Correspondences 1903–1955*. Traduction, Notes et Introduction de Pierre Speziali, Paris, Hermann

Fachinelli, E. (1983). *Claustrophilia*, Milano, Adelphi

——. (1992). *La Freccia Ferma: Tre Tentativi di Annullare Il Tempo*. Milano, Adelphi

Faimberg, H. (1988). The telescoping of generations: Genealogy of certain identifications. *Contemporary Psychoanalysis* 24: 99–118

Faulkner, W. (1931/1964). *The Sound and the Fury*, Harmondsworth, Middlesex, UK, Penguin Books

Ferro, A. (2002). Some implications of Bion's thought: the waking dream and narrative derivatives. *International Journal of Psychoanalysis* 83: 597–607

Fresco, N. (1984). Remembering the Unknown. *International Journal of Psychoanalysis* 11: 417–27

Freud, S. (1915). *The Unconscious*. In J. Strachey (ed. and trans.), *The Standard Edition of the Complete Psychological Works of Sigmund Freud* (S.E.) Vol. 14. UK, Vintage, The Hogarth Press and the Institute of Psycho Analysis

——. (1933). *New Introductory Lectures in Psychoanalysis* (S.E.) Vol. 22

Gallagher, S. (2007). Phenomenological approaches to consciousness, in M. Velmans and S. Schneider (eds), *The Blackwell Companion to Consciousness*, Oxford, Blackwell Publishing

Gieser, S. (2005). *The Innermost Kernel: Depth Psychology and Quantum Physics. Wolfgang Pauli's Dialogue with C.G. Jung*, New York, Springer

Grubrich-Simitis, I. (1984). From Concretism to Metaphor: Thoughts on some theoretical and technical aspects of the psychoanalytical work with children of Holocaust survivors. *Psychoanalytic Study of the Child* 39: 301–19

Hanly, C. (2007). Time and timelessness in psychoanalysis. *International psychoanalysis.net/wp-contact/uploads/2007/04*, consulted 23 June 2014

Jaques, E. (1982). *The Form of Time*, New York, Crane, Russak

Jaynes, J. (1996). *The Origins of Consciousness in the Breakdown of the Bicameral Mind*, London, Allen Lane

Johnson, M. (2007). *The Meaning of the Body: Aesthetics of Human Understanding*, Chicago, Illinois, and London, University of Chicago Press

Jung, C. G. (1934). *The Soul and Death*, CW Vol. 8, Sir H. Read, M. Fordam and G. Adler (eds), trans. J. V. Heurck, London, Routledge & Kegan Paul

——. (1952). *Synchronicity: An Acausal Connecting Principle*, CW Vol. 8, London, Routledge & Kegan Paul

——. (1974). *Letters, Volume II: 1951–1961*, ed. G. Adler, Princeton, New Jersey, Princeton University Press

——. (2008). *Children's Dreams*, eds L. Jung and M. Meyer-Grass, trans. E. Falzeder, Oxford and New York, Oxford University Press

Kestenberg, J. (1982). A metapsychological assessment based on an analysis of a survivor's child. In M.S. Bergman and M.E. Jucovy (eds), *Generations of the Holocaust*, New York, Basic Books

Kijak, M. and Funtowicz, S. (1982). The syndrome of the survivor of extreme situations: definitions, difficulties, hypotheses. *International Journal of Psychoanalysis* 9: 25–33

Lakoff, G. and Johnson, M. (1999). *Philosophy in the Flesh: The Embodied Mind and its Challenge to Western Thought*, New York, Basic Books

Laub, D. and Podell, D. (1995). Art and trauma. *International Journal of Psychoanalysis* 76: 991–1005

Laplanche, J. (1992). Psychoanalysis, time and translation. In J. Fletcher and M. Stanton (eds), *Jean Laplanche: Seduction, Translation, Drives*, London Institute of Contemporary Arts

Main, R. (2004). *The Rupture of Time: Synchronicity and Jung's Critique of Modern Western Culture*, Hove, UK, and New York, Brunner-Routledge

Matte Blanco, I. (1998). *Thinking, Feeling and Being*, London, Routledge

Penrose, R. (1995). *Shadows of the Mind*, London, Vintage

Primas, H. (2003). Time-entanglement between mind and matter. *Mind and Matter* 1(1): 81–119

Rayner, E. and Tuckett, D. (1995). *An Introduction to Matte-Blanco's Reformulation of the Freudian Unconscious and his Conceptualization of the Internal World*, London, Routledge

Rose, J. (1997). Distortions of time in the transference. *International Journal of Psychoanalysis* 78: 453–68

Sabbadini, A. (1989). Boundaries of timelessness: some thoughts about the temporal dimension of the psychoanalytical space. *International Journal of Psychoanalysis* 70: 305–13

Sartre, J-P. (1955). *Literary and Philosophical Essays*, London, Hutchinson and Co

Semprun, J. (1997). *Literature or life*, trans. L. Coverdale, New York, Viking

Sheets-Johnstone, M. (1999). *The Primacy of Movement*, Amsterdam, John Benjamins

Vinar, M. N. (2005). The specificity of torture as trauma: the human wilderness when words fail. *International Journal of Psychoanalysis* 86: 311–33

Wiesel, E. (1975). For some measure of humility. *Sh'ma* 5,100: 314–15

Wilgowicz, P. (1999). Listening psychoanalytically to the Shoah half a century on. *International Journal of Psychoanalysis* 80: 429–38

Winnicott, D. W. (1958). *Through Paediatrics to Psychoanalysis*, London, Tavistock Publications

Yiassemides, A. (2011). Chronos in synchronicity: manifestations of the psychoid reality. *Journal of Analytical Psychology* 56(4): 451–71

——. (2014/2016). *Time and Timelessness: Temporality in the Theory of Carl Jung*, London and New York, Routledge

2

SYNCHRONICITY IN CLINICAL PRACTICE

Caterina Vezzoli

Introduction

It all started in Sardinia, while on holiday with a dear friend, a Freudian analyst – a very good clinician with a solid theoretical background. One evening she asked me to listen to two dreams of one of her patients. Knowing how differently we approached the interpretations of dreams, I knew this was my friend's way to ask for my view that could help her gain more depth of perspective. Her patient was a very fragile traumatized woman. The dreams were dreamt in two different moments of analysis and involved a scorpion. In the first dream the menacing scorpion had sort of robotic movements; whereas months later, the scorpion of the second dream had lost the robotic quality and the patient knew she had the antidote, or was immunized against its poison.

The first association that came to my mind was that of the Egyptian Scorpion King of Pre-Dynastic and Early Dynastic Periods. This association was emphasized when my friend said that the scorpion of the first dream could be interpreted as the patient's dangerous father, overlaid by the patient's husband that had introduced her to the use of drugs. The Scorpion King is the unifier of the upper and lower kingdom: The patient, thanks to the analytic work, could deal with the poison by creating her own immunization. This was reassuring, if taken from the point of view of the prospective development of the patient. The containment offered by the analytic encounter allowed the patient's psychic energy to unravel, producing new symbols and letting the meaning emerge. The potentiality coming from the creative unconscious, previously blocked by the traumatic experiences related to the relationship to the father and to the husband, could now take the lead and develop further. The reference to the myth seemed useful as it offered a metaphor for the renewal of the 'kingdom' of personality. To deal with the scorpion, to be immunized, was the metaphor of finding a strategy to deal with the dangerous wounding father. These were my associations and my friend partly agreed with them.

My elaborations regarding the *potentiality of the unconscious* were difficult for her to comprehend, even though I had carefully avoided typically 'Jungian words' such as 'archetype' in an attempt to make myself clear; but to no avail. Suddenly a century of Jungian research opened in front of me: What could I say? Should I tell her about quantum physics or why myths are important? Should I speak about complexity and psychic system at the edge of chaos? Should I speak of the psychoid level? Of synchronicity? Should I speak of the use of amplification as a way of conveying meaning to the patient's material? Should I use the work of Bromberg (1998) on dissociation to try to introduce her to the Jungian way of thinking? Not on a summer night under the Sardinian sky turning purple, with the sea gleaming golden in the last light, and surfacing shadowy rocks that made me think that the world had just been created.

Validation of the theory

It was time for me to rethink the theoretical background that is taken for granted by my Jungian training; experience needed to be verified in light of my own experience and exposure to other theories. During the above discussion, I had in mind the importance of considering dissociation in relation to synchronistic phenomena, and I felt the need to clarify this point theoretically.

Long before Popper introduced his critique to psychoanalysis of not being a scientific discipline as its methodology was lacking the possibility to falsify its hypothesis, Jung had set up experiments that could verify the repression mechanism proposed by Freud. The World Association Test (WAT) sprung from Jung's interest to anchor psychology and psychoanalysis in the ground of scientific disciplines. Proceeding in his scientific research on WAT he worked at the root of human psychological functioning and came to verify and elaborate further the notion of complexes: the feeling tone around which the experiences of the subject aggregated. In the experimental research on WAT Jung found that the feeling tone was made up of early relational experiences that were lived in the relationship with the care giver, and were channelled by a model of relational functioning present as an innate pattern, the archetype.

As early as 1912, Jung started to question the concept of libidinal energy and defined energy, as Hester Solomon summarizes in this quotation:

> The psychic manifestations of this generalized energy take the form of images, and these in turn become symbols when they act as transformers of energy, eventually offering the possibility of new resolutions of old problems. We know that the archetypal image can reside both at the personal and at the collective level.
>
> (Solomon 1991: 307)

The concepts of archetypes and the collective unconscious were defined in Jung's paper '*Instinct and the Unconscious*' (Jung 1919). The definition of archetype that I consider most valid states that archetypes are potentially present in the psyche as patterns of behaviour that are activated during different events or during a period of one's life; the degree of activation depends on the dynamic between archetype and complex. In the unconscious, the archetype is present as *potential*: in order for it to emerge, certain conditions are required. Implicit and explicit memories are stored and aggregated in the complexes according to a pattern of behaviour that, when activated in the analytic relationship, can be investigated as part of the transference–countertransference relationship.

The potentially undifferentiated archetype is incorporated into the complex where the experiences of life events, early relationships (positive, negative or traumatic) are stored. The archetype is the element around which the complex aggregates feelings. In the dynamic between complexes and everyday life, aspects of the archetype may emerge in the form of symbol, and convey new meanings. We do not always have symbols; often we have pathology, symptoms, and dissociation. The energy of the archetype is responsible for synchronic phenomena and for the emergence of unexpected images, meaning and symbols that might resolve the contradictions of a situation.

Analytical psychology and new physics

In the years of the great discoveries of Einstein, Pauli, Heisenberg and Bohr, Jung considered that psychology had to become a science – or, better yet, '*the mother of all sciences*' (Shamdasani 2003: 93). With great openness of mind, Jung confronted his developing theory to the modern theory of Physics. He tried to understand the personal equation in the light of the new paradigms science was elaborating. He was looking for a unifying theory of science that could include psychology. Many years later, in one of his last letters, Jung stated that what defines a scientific theory is its heuristic value and its applicability to understand nature: '*Psyche is the mother of all our attempts to understand nature*' (Shamdasani 2003: 99).

Today, scientists agree that scientific models must have a heuristic value and that theory and practice should partake in the construction of scientific knowledge; there was no such consensus in Jung's time. However, at the time the scientists studying the new physics were arguing that the reality of the world of physics has nothing to do with the reality we perceive through our senses. This viewpoint was in contrast with the classical physics of Enlightenment that considered scientific observation to be describing 'reality'. Reality in the world of quantum physics is in the realm of *potentials*: the particles come into being when energy is given to the system; they are not present before. Nothing happens without an observer. As Yiassemides (2014) emphasizes, in the unified field of the archetype the observer unifies conscious and unconscious, and attributes significance; *internal causality is determined by the observer's perception.*

Anna's dream

Dreams come from a dissociated state of mind and constitute a way of dealing with traumatic events. Anna, one of my patients, a young woman, dreams of a house at the seashore where she is looking at a small tornado that is developing in front of her while she is watching: 'I'm in a house with a view at 180 degrees on the sea. From the sea rises a vortex, a small tornado. The house is a Mediterranean house with white columns. The house is semi-circular and at ground floor. You can't see the seashore that is just underneath the house, you can see rocks, small rocks. From somewhere my mother enters the house, maybe she has shopping bags; the view of the small tornado that moves at the horizon is amazing.'

She tells me that the vortex rises from the water: 'It is as if the sky is extracting the water from the sea. It is a sort of primordial image.'

I'm surprised by the description of the image. The previous month, for the first time in my life, I had witnessed the phenomenon from my house at the seaside; a French colleague and friend had sent me the photographs of the phenomenon that I had never seen before. My patient's description makes me think of the Greek myth of creation, where the masculine and the feminine principle violently collide to create the world.

Anna tells me about the differences between the house of the dream and the house her family owns at the seaside. While my patient is describing her associations to the dream I ask myself what it is *really* happening – why is the dream so important? Just then, Anna offers her explanation: 'It's the first time I dream of a tornado; usually I dream of earthquakes as I already told you.' In the Italian region my patient comes from earthquakes can happen. I ask her what the difference is between earthquakes and tornados: 'In the earthquake there is a violent shaking of the ground while in my dream, sky, wind, and sea attract each other, they get violently fused and the water of the sea is sucked into the sky. The sky is big and the sea rises towards the sky; it reaches towards the sky.'

I think to myself that this is an early Creation Myth: the violence of the masculine on the feminine principle. I ask in a soft and warm voice, feeling very near, very empathic: 'What do you think tornado and earthquake are telling us?'

Anna answers after a while: 'I think the earthquake represents the upsetting at the beginning of my life. In our session a few days ago when I was speaking about my difficulties in relationships I told you about the earthquake I feel inside; you stopped me on the word "earthquake" and I spoke about the upsets that are at the origin of my personal life, and that of my family.'

I had followed her along the path of her associations that centered around her first years of life, on how the family chronicle reported by her father were tales about her not being properly fed by her mother. She wondered about the truth of what her father said: Was it a dramatization or was it reality? She had memories of when she was four or five years old of her little friends saying that she was smelly. She concluded that she was not very clean. When her grandmother would visit, she would give her a bath, change her underwear, and put on a new dress. Many years later, during her analysis, my patient's explanation of her mother's behaviour was that she was probably severely depressed, but this was not mentioned in the family chronicles. She remembered that when she was a teenager her mother had been treated for depression.

Parallel process as mirroring

Anna is wondering about her first years of life and I suddenly realize that she is not breathing! She is holding her breath, as if she cannot breathe, and I realize that *I am doing the same*. I say to myself 'I'm mirroring her; I'm attuned to her feelings.' Then I remember *my* family chronicles: When I was a child of about two years of age I had fallen ill with a severe illness that in the fifties was considered dangerous and treatable only with newly discovered antibiotics. Both my parents reported their concern about my illness especially my mother that for the subsequent fifty years would talk about her worries at the time and emphasized how healthy I had been ever since.

Anna says: 'I believe that the possible meaning of the tornado is to be taken into the vortex, suspended between the sea and the sky; but in the dream I am secure in the house.' To which I reply: 'the Titans are outside; it is a battle of titanic forces you are witnessing.' And then I ask: 'And what about the mother of the dream?' After a silence Anna replies: 'I don't know. She has nothing to do with the house, maybe she is carrying shopping bags.' She goes on saying: 'When I was very little my mother's milk wasn't nurturing, I didn't grow, my father says he was the one that saved me, he says: "I have fished you out." My older sister, three years my elder, was jealous of me and they had to look after her because she did things to me . . . I was a bit squeezed by everybody.'

I think to myself 'No wonder she has difficulties breathing.' As I am wishing that she could breathe more easily, I ask if the house of the dream is secure enough. Anna answers: 'There are no dangers, it is a secure house.' A long breath on her part ends the exchange. Again, my association goes to my house on the beach when there are sea storms, and how secure I feel inside.

Trauma and dissociation

The dream opened a scenario of hidden traumatic memories; it showed a need to uncover the dissociated memories related to the trauma, and to find possible symbolization. Traumas and early relational experiences are recorded by the amygdala. The emotions of the first days and months of life are registered by the amygdala. Neuron-research on dreams and brain suggest that the amygdala may influence the REM sleep and the production of dreams through the process of emotional memory previously stored. Now we go back to the process of formation of complexes. Memories are stored in the complexes; experiences of our life since the beginning, implicit and explicit memories are both present. The archetypal core of the complex constitutes the potentiality that will be made explicit by the interaction between spirit and matter, between memories and experience, present requests and reflections, and transformation of past and present dimensions. The archetypal core constitutes the non-differentiated matrix of a possible output that registered in the substrata of the complex. To develop in an actual form, it must be contained and given attention to, both intra-psychically and in the outer relationship.

The patient's sub-cortical system produced a dream that her cortical system translated into symbols. The coherent narrative she was trying to produce in the session was a way of declining the early emotions within the analytic relationship that could contain the energy of the archetype; an energy that could otherwise take a destructive direction. This is my explanation as to why Anna noticed that for the first time she was dreaming a tornado and not an earthquake, and described the difference of feeling between an earthquake and a tornado or a vortex.

The image of a tornado had to do with being violently extracted and absorbed into something else; whereas in the earthquake there is the trembling and the opening of a chasm and of sinking in an abyss without end. Both experiences are frightening: as mentioned, they recall the creation myths where the feminine and the masculine principle violently collide to create the world. Anna felt that the vortex was extracting something that would otherwise remain submerged; it was better than to sink into the abyss without any possibility of light. She went back to the threatening beginning of her life where she was in danger of falling into the abyss of death. My complexes responded to the power of the archetypal image produced by her and I regressed to an experience that was similar to hers.

The psychoid level

Anna and I were both in an archetypal present and we responded by re-processing our experiences and finding a less violent pattern of relationship. We could cry about her depressed mother and dramatic father and empathize with the little child that was going to die and had to be helped into life. All this should be seen metaphorically: *near death experiences are both concrete and spiritual.* My introduction of myth was a way of responding to the dissociation, an attempt to bridge the split that had occurred inside her and my psyche, using amplification instead of interpretation. When I wished for her to breathe I was trying to help her out of the psychoid level which was localized in the body, so that she could move from the pure body to the level of being in a body-state that can be recognized by the mind. I was not only mirroring her; we were *both* in the psychoid field. My enactment was an attuned response to her distress.

In the session, we experienced the time as Kairos, the supreme moment, as the opportunity of colliding and disintegrating space and time, where separation of past, present and future does not exist. From this point of view, what happened in the session was a *synchronistic event*. Synchronicity is a form of transcendent function, in the sense that it creates a synthesis between two oppositions. The rupture of time that had taken place had brought us at a situation deep down in our bodies, where memory was confused with our species' instincts, where mind and body were objectively one. The potential images of life and death were not accessible, so we needed an image coming from the myth – humanity's cultural bootstrapping – that could give meaning to the personal experience we were living.

Enactment as amplification

Since the psychoid level was not yet declined as a potential, my introduction of myth was a way of making a connection between the psychoid and the psychic levels, through amplification. An interpretation would have created a further rupture. Amplification, by connecting the images of the dream to the metaphor already present in the associations of the patient, created a transpersonal narrative that was needed to bridge spirit and matter.

In the situation of unformulated unconscious, 'by evoking, through enactment, perceptual experience that doesn't fit in [we allow] narrative change to take place' (Blomberg 1998: 176). The technique of amplification is about narrative; it is the Jungian way of helping new meaning to emerge. This theoretical reflection is related to Cambray's hypothesis that Jung used amplification to generate symbols that would 'engender awareness of interactive fields and emergent states of mind' (Cambray and Carter 2004: 233).

Lakmé and Malika

The second clinical vignette starts with the evocation of a musical image without sound, and is followed by a synchronic event that modifies the transference and countertransference issues. Malika started the session by saying: 'You know Doctor, I discovered that I am a dramatic soprano.' The quality of her communication is, dare I say, happy; her expression an open one. I myself feel relief, recognition, and perhaps joyful freedom. I briefly mention my countertransference reception of her communication and shared feelings. To which she adds: 'In a way I've always known myself to be a dramatic soprano, I was attracted by roles such as Tosca in "vissi d'arte, vissi d'amore", but my teachers had always said that I could forget about it, it was not for me. Do you remember when some sessions ago . . . about my dream . . . you asked me how I could describe my feeling and I talked about the duet in the Lakmè of Delibes? Do you remember when the two women descend along the river with the boat and contemplate the luxuriant shore rich in plants and flowers singing to the beauty of life?'

I had only a vague idea of the duet she was talking about: an old memory of something I had been obliged to listen to and was, therefore, a bit annoying. I knew something about the story but I did not try to recall it, not even when I understood that for my patient this Opera had meaning, my position being that *if this is important the unconscious will find a way to use it.* However, I recalled in detail the previous session where I had invited the patient to further associate on an image of a spontaneous association she had reported in reference to a dream.

Malika continued: 'After our session I tried with my new teacher to sing the part of the dramatic soprano in the duet, to my surprise it not only corresponds to the characteristics of my voice but I was very deeply into it. My previous maestri . . .' Her question hangs in the air unanswered and is a question for both of us. I say tentatively (but it is not the only thought or feeling I am experiencing): 'Your teachers were probably wrong but perhaps your dramatic temperament was inhibited, perhaps you were too committed to being good, reliable, and secure for others, for your husband, your parents . . .' The atmosphere between us changed: lightness disappeared, and suffering entered the session. I wondered if what I had said was too concrete, if I had not broken the communication of empathetic feelings. However, the tone of my voice, the way of looking at her, what I felt while speaking to her, was as if I was caressing her; I hoped I had transmitted these feelings to her.

Then I had the perception that both of us were remembering the difficult years of her childhood, her loneliness, her mother's cruelty, and her father's absence; the 'saving missions' she embarked upon while in frustrating relationships with the wrong partners. How could she afford to be 'dramatic'? It would have meant living the traumas anew. Malika's salvation was to deny the drama and live 'as if.' She felt she had to be a mezzosoprano, *as if* her voice was not pure enough. She had never said it clearly, and not knowing the '*bel canto*' I was not sure of the meaning. I was under the impression that the mezzosoprano or contralto was played by a young man that had been castrated to maintain his feminine voice. I was not sure of this; I had seen the film 'Farinelli' and had learned, during my studies of Medieval history, that for many centuries women could not speak loudly in churches. These vague thoughts made me perceive the great loss my patient was suffering, imprisoned in a voice or maybe in a skin that was not hers.

After a silence charged with emotion she said with moistened eyes: 'I wanted to be good at all costs and I caused disasters; only now at this point of my life can I afford to be a dramatic soprano because I am starting to live *my life*. What I long for is to be able to give my voice to the heroines of dramatic Operas!'

The analyst and synchronicity

Days later, on a Sunday, I found myself in my car, on my way to have lunch at my cousins' in the countryside. There was nothing worth listening to on my usual radio programme; I looked for another radio station and I found a French broadcaster where two women were singing. While listening to the duet I was struck by the sweetness and dreamy quality of the voices of the two women which were intertwined while expressing their diversity. I thought '*a dream-like quality*'. It wasn't until the presenter announced it at the end that I realized that I was listening to the Lakmé duet; the same duet my patient was talking about.

I questioned whether this was synchronicity or coincidence. But mainly I wondered what 'a dream-like quality' meant. Could it possibly be related to an *intuitive empathetic feeling*? Going back to the exact thoughts, feelings, and images the few seconds between the singing and the broadcaster's declaration of the title, what I remember is the appearance in the distance of the cupola of the church of my infancy and me thinking: 'I'm nearly home.'

Operas and alienation

What could be the meaning to Malika of the two women in the boat, descending along the river, looking at the beautiful riverbank passing by, and exchanging feelings and beauty? Were they two lovers, two friends, mother and daughter, analyst and patient, teacher and student? Eros, who sprung from their singing and their voices, celebrated life. I recalled the shared moment we had built in the previous session, and what we had felt together. The voices were a reminder of the empathetic linkage between analyst and patient. The duet was a perfect representation of the relationship that we were co-constructing.

Why Opera? I am ignorant when it comes to operas; I know very little and I can hardly distinguish between a soprano and a mezzo-soprano voice. However, when I was listening to the music in my car I could hear the interconnection and correspondence of the voices. I am not able to grasp nuances (even though when I was studying at University I had been part of the La Scala Clique). I was never among those who wanted to go to the Opera; I found them terribly boring; I preferred classical music concerts, or the ballets. When I was young, the Operas at La Scala were for the bourgeoisie. And so, I remained ignorant.

Nonetheless, in my analytical training I was lucky enough to work in a seminar on Mozart's *Magic Flute*, where I learned to recognize and associate changes in the voice and quality of feelings. I learned to pay attention to changes conveyed by the voice, to understand how in Opera the music and the different voices are underlying the feelings that, through emphasis placed on the *canto*, reach expressive qualities. The words used are not as important as the voice; it is *how* the words are sung that express the nuances of feeling. Thus, I could not ignore the message that came through the voices singing the Lakmé *Flower Duet*.

Singing, a world unknown to me, was my patient's way of communicating her distress. Malika said, 'I have the extension of voice to be able to sing that part, but something inside me shuts down, and the voice does not come out. It is useless to try to make an effort because it makes it worse. Even the Maestro's teachings are useless. It is something that I feel here . . .' – she indicated the centre of her chest. The sorrow, the sense of alienating inability she was expressing while expressing her feelings, her experiences, allowed me to understand empathetically her loneliness, her frustration, her need to be able to break the spells that kept her segregated in the prison of '*disaggregated affects*'. At first, my ignorance kept me alienated from the inner world of my patient but the synchronistic event allowed me to listen without prejudice, and to discover her unknown world which I had thus far overlooked.

Captivated by the mystery of voices, I could at last understand how important singing was for her. Implicitly, she had known that she had the capacity to sing her drama. She had within herself the potential, but for a long time an obscure force had blocked her voice. She was now able to appropriately express the range of her distress.

Transference and countertransference

I did not know the work involved in the expression of feeling through the voice when interpreting Opera. Nonetheless, at some point I became empathetically tuned into a feeling that desperately sought to be expressed through the voice that vibrates inside the one who produces it, and reverberates to evoke the same feelings in others. I could feel that singing was my patient's way of alleviating her early traumas engraved in her procedural memory. Singing her trauma meant recalling, in order to dissolve it, while experiencing in voice the different nuances of its content. Describing her sorrow for being unable to express the dramatic nuances of her experiences through singing opened the pathway towards her authenticity, and allowed me to accompany her in the regression that was necessary to unravel the traumatic connections.

Singing was the preferred tool in Malika's world; in her solitary, difficult, and devastating childhood, singing was the only thing that belonged to her. Singing, a world so elusive to me, was present in all its complexity in our relationship. At times, it seemed that I understood what was happening; other times I was in total darkness but still confident that our relationship was maintained at a deeper level.

Before the Duet of Lakmé emerged in the force of an aggregating symbol, I was working around co-construction and implicit learning. I am referring here to the work of Beatrice Beebe and Frank Lachmann (2002) on 'implicit relational learning'. The concept describes the constituents of the relationship that catalyse the formation of representations. The ability to internalize comes from the way in which mother and child (and analyst) continually build, develop, and represent the emotional adjustment, which is simultaneously interactive and self-regulating. Internalization is possible through the experience of co-constructing in the relationship.

From neuroscience and research on mirror neurons we have acquired the knowledge of a genetic predisposition to co-construct the relationship. We have the capacity to read *intentions*: In the brain of primates and humans there are neurons able to assign an intention to the actions of others, and to the relationship with others. From the field of analysis, and particularly from infant observation, we know that the exchange between mother and infant are intense and that the infant responds on a selective basis to maternal communication. Winnicott has also explained the importance of *maternal reverie*. From Fordham's work, we understand that the ego complex is built and developed through the relationship and the processes of integration and disintegration; similarly, the other complexes such as the shadow and the personality are built.

In search of meaning

Why did the Duet emerge so strongly in me, and why did it have such meaning? The synchronistic event in my car had the purpose of calling my attention to the importance of voices in the expression of feelings. Singing is a structuring aspect of my patient's personality; the nuances of her being and feeling are expressed *through* singing. Her difficulties in singing are the metaphorical aspect of her difficulties in psychic development. Malika's voice blockage could really be attributed to an emotional implicit interference coming from her body, interfering with her expression. My patient's intuition was moving along these lines. My own intuition, following hers, was that singing – by transposing in the voice aspects of feeling – would allow for the manifestation of a pre-reflective aspect; through singing an awareness of feelings removed at an embryonic state would be allowed back through the transcendent function.

In due time, the creation of a new symbol that will go beyond the conflict will appear, creating a new possibility in the well-known dynamic of the dialectical functioning. The discovery of having a voice that could reach the extension of the soprano's range could be considered as possessing a pre-symbolic function, situated between the reflective and the transcendent function, as an attempt of the psyche's self-organization to find a solution to traumatic events. In further analysing with my patient what it meant to her being a soprano or having a soprano voice, she explained that when the voice reaches its highest notes she would feel a sort of upheaval vibrating in the air. Once she had reached the top, a trace remained: the wish to free again the voice, as well as the feeling associated with it; the wish to express something deep inside: love and despair, despair of love.

The psychoid

Regardless, *the experience* of singing had not yet been responded to adequately. The image of that duet, which symbolized our relationship, needed a synchronistic event to attract my full attention. In which way was the event on that Sunday morning in my car related to me, to my feelings, to the relationship to the patient? Thinking in terms of the psychoid layer of the psyche, what remained to be explored was the level of my interconnection to the world of my patient.

Cambray (2009) describes the concept *psychoid* as the quality of the mind that is the closest to the biological state, and it refers to the feature that some physical and sensorial perceptions have, in that they cannot be physically represented. These perceptions do, however, contain potential resources and possibilities that are unknown to consciousness. Psychoid is therefore 'quasi-psychic' and it acts as an interface between the psychological level and the material level, which are, yet, undifferentiated and therefore not perceivable by consciousness.

When I explored the event, trying to find the inner meaning for myself, I recognized feelings that were unknown to me: as I was going to my cousins' for Sunday lunch I was returning to my infancy, to the heart of my family; I share with them my oldest memories and earliest relational experiences. Now that our parents have died, to meet is to bring together what we are with our ancestors; it is a way of symbolically reconnecting to the common land that gave birth to us.

The landscape has changed a lot through the years, especially now that the new highway has destroyed many landmarks of our childhood and made 'our land' unrecognizable. We even joke about the absence of familiar reference points: 'You know I couldn't recognize our uncle John's farm because it has disappeared among the new highway bridges; so I got lost coming here . . .' However, the cupola of the church is still there, and it is still recognizable from a distance, reassuring us that we are still remembered, somewhere. The Father God of my childhood is still there waiting. I do not practise religion anymore but there is something inside me that says: 'You are welcome to this village. It's your place.' To emphasize the importance of this landmark I should mention that my first dream when I started analysis took place in the churchyard of this very village. As a child, I used to go to the churchyard with my parents who told me the stories of our ancestors; I have known, since then, that in the end I will return to be buried in the family chapel of this same churchyard. For me, the church and the churchyard are the places that symbolize life: the beginning and the end.

We could normally describe my thought as my own regression; what the synchronistic event changes via the emergence of the Duet is the connection to my patient. Through singing she opened her world to me and I understood that the common ground we share are those early feelings. I had to follow her in the undifferentiated world of the psychoid where we could reach the voices of women, feelings beyond pathology; to reach into the soothing sounds of women's voices.

Conclusions

The setting of therapy spread from the analytic room to the analyst's enactment, facilitated by the material of the patients' dreams, in an attempt to bridge the split created by the dissociative traumatic memories. In both cases, the traumatic experiences emerged in dreams, and entailed synchronistic events. In both patients, early traumatic experiences presented aspects and memories of disorganized attachment, that could justify the dissociation, necessary for both of them in order to survive the early ordeal.

For Jung, complexes can be autonomous and the psyche tends to split; when a conflict is unbearable, the psyche dissociates. When a complex dissociates, part of the complex can become autonomous and create a 'splinter psyche' that can overcome the entire personality, while in a normal situation the Ego can maintain its function. As explained above, in the complexes there is an aggregation of memories and experiences that can be activated at different moments, depending on life events; life experiences are received differently depending on which parts of the complex are being activated or constellated. There is a plurality of shades that constitute each complex as well as several levels of consciousness; the healthy way to deal with the different levels of a complex is to maintain a link with the Ego complex.

The psyche in Jungian terms is not a unity, and 'normal states' of dissociation are possible. Complexes are always partly unconscious, and, in the area of implicit memories, unconscious contents that the ego cannot integrate (due to lack of understanding or lack of attention) are stored. As the process of creation of complexes starts at the beginning of life in interaction with the caregiver, the very early experiences can remain incomprehensible or go unnoticed. When a life event 'strikes' these areas, the unconscious reacts by dissociating to primitive mythological or instinctual levels in order to avoid the incongruousness that arises by unconsciously activated contents that are incompatible with consciousness.

The psyche is a multi-layered structure and trauma severs as the psychic connection between layers. In my patients, the dissociated aspects were evoked by the dreams and it was by working on the regressed aspects brought about by the images and contents of dreams that we could emotionally reach the dissociated elements. During the regression that took place it was possible to contact images and aspects of the collective unconscious (like for example the images of the Titans) and the layer of the psychoid where matter and psyche form a continuum; this is the space where synchronic events can take place.

Traumatic experiences, at some level and differing intensity, are present in all of us and determine our fragile aspects that predispose us to regression and dissociation. My patients' traumas concerned the early interaction with their caregivers and the lack of reassurance and care. The right to exist was not granted to them. Anna was lacking fundamental care that would assure survival; and Malika was robbed of the possibility to love. Their primary needs – the right to be recognized, loved and accepted – were expressed differently and evoked different reactions in the transference and countertransference fields. The synchronic event with Anna was an image of the dream followed by a body-state that allowed a recognition and expression of feelings; the analyst's enactment contained the experience and helped to transform it by giving it meaning. The enactment was an automatic action in response to an urge to take care, that was received as recognition of a forgotten need for mirroring.

With Malika, the synchronic event in the life of the analyst happened in response to a loss of early investments in life, the sense of alienation and unrelatedness that can accompany the destruction of early love investments: My synchronistic event was an expression of projective identification that had to do with the lost world and a longing for a return to Mother Nature. I, the analyst, had to work on my dissociated aspects of loss and longing in order to reverberate the importance of the voice of a dramatic soprano.

In both cases, synchronicity represented a sort of *therapeutic strategy* that created a bridge in the inter-subjective field, so that the dissociation could be transformed in a meaningful event; as Jung said, 'dissociation and transmutation are in relation to each other' (Jung 1935/1976: 139).

References

Beebe, B., and Lachmann, F. (2002). *Infant Research and Adult Treatment: Co-constructing Interactions*, Hillsdale, New Jersey and London, UK, Analytical Press

Blomberg, P. M. (1998). *Standing in the Space. Essay on Clinical Process Trauma and Dissociation*, Hillsdale, New York, Analytic Press

Cambray, J. (2009). *Synchronicity: Nature and Psyche in an Interconnected Universe*, College Station, Texas, Texas A & M University Press

Cambray, J. and Carter, L. (2004). *Analytical Psychology. Contemporary Perspectives in Jungian Analysis*, Hove, UK, and New York, Brunner-Routledge

Deacon, T. W. (1997). *The symbolic Species. The Co-Evolution of Language and the Brain*, New York and London, W.W. Norton and Company

Delibes, L. (1883). Flower Duet, *Lakmé*

Dennett, D. (1993). *Consciousness Explained*, London, Penguin Books

——. (1996). *Darwin's Dangerous Idea*, London, Penguin Books

Hendriks-Jansen, H. (1996). *Catching Ourselves in the Act. Situated Activity, Interactive Emergence, Evolution, and Human Thought*, Cambridge, Massachusetts, MIT Press

Jung, C. G. (1919). Instinct and the Unconscious, in *The Structure and Dynamics of the Psyche*, CW Vol. 8, Sir H. Read, M. Fordam and G. Adler (eds), trans. J. V. Heurck, Princeton, New Jersey, Princeton University Press

——. (1935/1976). *Determinanti psicologiche del comportamento umano*, tr. it. in Opere, CW Vol. 8, Turin, Italy, Boringhieri

Kalsched, D. (1996). *The Inner World of Trauma. Archetypal Defenses of the Personal Spirit*, London and New York, Routledge

Pauli, W. (1994). *Writings on Physics and Philosophy*, Berlin, Springer

Shamdasani, S. (2003). *Jung and the Making of Modern Psychology. The Dream of a Science*, Cambridge, UK, Cambridge University Press

Solomon McFarland, H. (1991). Archetypal psychology and object relations theory: history and communalities. *Journal of Analytical Psychology* 36: 307–29

Wilkinson, M. (2006). *Coming into Mind*, Hove, UK, and New York, Routledge

Yiassemides, A. (2014). *Time and Timelessness: Temporality in the Theory of Carl Jung*, London and New York, Routledge

3

TEMPORALITY AND THE TORMENTS OF TIME[1]

Ladson Hinton

Homo Temporalis

Our temporal nature is also our tormentor. Murky shades of the past and future haunt us. As William Faulkner observed, 'The past is never really dead. It's not even past' (Faulkner 1951). We could accurately dub the human being as *Homo Temporalis*. Everything that we experience has a temporal dimension, and temporality is at the core of subjectivity, especially the process of remembering. Time drives us, inspiring both our nobility of purpose and our destructiveness, and it is strangely elusive, appearing to vanish as soon as we try to describe it.

'Space contains [both] living and inert bodies; however, only the living human, hence the living psyche, is subjectively concerned with time' (Scarfone 2006: 810). Entering time involves loss of the illusion of safety and unity. We enter the world helpless, immersed in an enigmatic flood of messages, faced with the task of becoming subjects through the establishment of memories that anchor our sense of self. Thus, though entering time frequently involves an experience of disruption and loss, it also gives us substance and reality.

The cave art found in Southwest France powerfully evokes the potency of time. I was fascinated by it for years, and finally visited several sites during the 1980s. There are hand impressions on many of the cave walls, purposely done in realistic outline as if spray-painted. I had to restrain myself from touching my hand to theirs – alarming the guides when I made a gesture in that direction! The hands seemed to reach out directly to me, their visitor of the future. My feeling of kinship with those Cro-Magnons of 30,000 or more years ago was moving and powerful. These experiences increased my wonder about time and memory as the quintessential human traits.

The vastness of the caves, sometimes narrowing into passages that force you to bend and squeeze through, invokes primal fear and fascination, an attitude of both anxiety and surrender. The surrounding pressure of the earth and the sense of an underworld, of another world, are pervasive. To create and then ritually return to such paintings deep in the caves, with only tiny lamps and handheld torches to provide light, required extraordinary motivation. A strong sense of spiritual and psychological renewal must have drawn these ancient people to these recollections that were so difficult to access (Leroi-Gourhan 1967; Campbell 1983). The artists and their comrades must have possessed an awareness of the future, imagining times when they would return to view their handiwork for ritual or other reasons. Like all artists, they probably envisioned others in an unknown future also regarding their works.[2]

One might speculate that ritual visits to such paintings manifested a drive to disrupt the grip of repetition (*agieren*), of circular action and reaction, the mere going on being of everyday life (Marucco 2007; Scarfone 2011). Remembering, continually re-creating the past, is the living ground of our subjectivity. Indeed, mind and memory are etymologically related (Loewald 1972: 408). 'While repetition constitutes the basic level of mental functioning, remembering . . . [is] a fragile, pulsating, discontinuous, almost evanescent feature . . . [that involves] recomposing one's whole mind' (Scarfone 2011: 71). In this view, remembering is an activity of differentiation, whereas uniformity and circular movement are typical of repetition, in which nothing happens but identical 'nows' (Merleau-Ponty 2012: 167). Differentiation generates anxiety because it provokes a fear of fragmentation and consequent loss of an imaginary unity.

Temporality, then, is basic to reflective thought and the creation of culture and, particularly, to the theory and practice of psychoanalysis (Loewald 1972: 401). Jung did not elaborate a theory of time in a single work, but some of his comments convey a multifaceted view that was somewhat innovative in its day. I have critiqued Jung's views on time elsewhere (Hinton 2014a).[3] His considerations about remembering ultimately depend on the assertion of an archetypal reality that 'transforms the past in the service of wholeness' (Jung 1930: para. 54). However, reliance on 'archetypes' and a teleological view of experience can minimize concrete experiences of the past, including remembering, in all their substance. In fact, we are stuck with whatever our own time's images and gods happen to be, and we must always begin there (Giegerich 2001: 77).

Most of the recent writings in the Jungian literature regarding time focus on synchronicity. Von Franz (1974) made early efforts to explore the meaning of time, and Yiassemides (2014) recently did an excellent job of pulling together Jung's basic thoughts on the subject. In contrast to those works, my own concern is 'the time of our lives' and not the 'time of the universe'. I will use 'time' and 'temporality' somewhat interchangeably, although my basic concern is 'temporality': the ordinary experience of our analytical work and our everyday existence. In my opinion, such temporal experience precedes the universal time of science or archetypal theory and not vice versa. We are present as temporal creatures in our own lives before we speculate about the cosmos (Hinton *et al.* 2011; Brooks 2011; Hoy 2012: xiii–xiv and 92–93; Hinton 2014a).

From the ethical point of view, our personal and cultural pasts are written in time and we are responsible for our acts, although we may interpret and reinterpret their meaning (Caruth 1996). If one privileges the archetype as retranslating the past (i.e., via *Zurückphantasieren*) it would seem to diminish a feeling of concrete ethical responsibility. For instance, it lends itself to the view that 'it was archetypal,' detracting from a sense of personal or collective accountability. In Levinasian terms, the concrete 'face of the other' can become obscured by abstraction (Lin 2013). The practice of social ethics and justice are strongly affected by our views of temporality (Hinton 2014b).

Temporality in analysis: the case of Ralph

After this brief introduction to the labyrinths of temporal horizons, I will relate a clinical case that provides a concrete sense of the workings of temporality in analytic work. That will provide the groundwork for further reflections. I will conclude by discussing some recent findings in the neuroscience of time that show a surprising degree of resonance with contemporary psychoanalytic views. The following narrative chronicles the violent breakdown of a lifelong state of being partly 'frozen' in time, walled off from the flow of life. It describes how traumatic disruptions can open the possibility of a more dimensional sense of time, and conveys in depth the transition from a defensive state of iteration to one of alteration, and to a more open temporality.[4] I thank my patient, 'Ralph', for being willing to share these experiences.

Ralph was in his mid-fifties when I first saw him. At our initial meeting, he was in anguish, almost to the point of panic, and quickly broke down in sobs. He had been without sleep and emotionally desperate for several days. Between eruptions of emotion, he began to tell his story.

His present turmoil had begun after the breakup of a brief love affair with a woman near his age. It had been the most intense and passionate experience of his life. As events evolved, he could not accept being separate from her, insisting that they be together constantly. 'All I can think about is her!' he said. As time progressed, she had apparently found the situation to be overpowering and, perhaps, frightening, and had called a halt, refusing to see him. This rejection was a tremendous shock, and he was shattered and in despair, constantly beseeching her for contact – to no avail. She and his friends told him that he 'needed therapy'!

His wife of many years had died one and one-half years previously, and this relationship was his first since her death. He described his former wife as a 'good' person to whom he had basically abdicated the raising of their three children and their interpersonal world. She 'ran their lives'. Ralph was very talented in math and science and had been successful in biotechnology.

We began to meet three times per week, and quickly formed a distant though friendly working relationship. I noted that he increasingly watched me carefully to see the way he was being perceived, but it also came through that he was very motivated to understand the state of his life. He did not come across as merely compliant, although that had often been one of his tendencies in the past. Occasionally, there were sparks of humour. I commented very little and primarily listened, asking occasional questions.

He began to discuss the pattern of his limited number of relationships with women. Most frequently, he had devoted his time and energy to any woman who appeared to provide acceptance and direction. 'I would give up my very being to be loved,' he would later say. This type of relationship had largely been his experience of marriage, and he thought that he had remained emotionally 'frozen' as a husband and father, although he functioned well in his career.

During the following weeks, the overwhelming emotions settled down with the containment of the analytic setting, and he appeared to progress with intellectual insights regarding his past and present situation, particularly his profound dependence on women. He had a scarcity of memories, and when he spoke of past events, I strained to visualise the interpersonal context. It was like hearing the report of a visiting social worker. When he visited his sister and her husband in another city I had no sense of their or his presence or interaction. It had the quality of uncanny absence rather than mere coldness. This tendency was present with most of his early memories, until the death of his wife and the recent affair.

The most typical focus in the past was his 'cold, disconnected' mother. She was a certified accountant who maintained a busy practice during most of his childhood. He reported that he had no memory of her being affectionate with him or his sister. However, he vividly remembered photographs of his mother as a young woman who appeared to be happy and vivacious. These pictures haunted him. How could that be the mother whom he knew? Why had he never experienced such vivacity from her? His childhood had been filled with a sense of lack and dread, a feeling that others had access to something he was denied. Gifted in math and science, he took some pleasure in his academic achievements from an early age. He rarely spoke of everyday past events, or of 'chums' from his early years, and seldom mentioned friends from any era of his life. A tangible sense of pastness seemed almost absent.

After approximately two years a new focus suddenly emerged. Ralph entered the room and sat down, intently staring at me with a fixed but expressionless gaze. I immediately experienced a powerful set of emotions. They were inchoate, 'other' – nearly impossible to capture in words. It was like a wave of dark blankness swirling in my mind, and I was stunned. After a few moments, I looked at him quizzically; however, he was somewhere else in his mind and not really seeing me. It took all my fortitude to maintain my capacity for reverie. After several minutes his visage shifted and he quickly continued with the details of his ongoing life as if nothing had happened. When I asked, 'What was that?' he just shook his head with a grimace of extreme discomfort, saying that it was a terrible feeling and that it was a horrible place he had *always* known. He had no words to describe it. As we talked, he became extremely anxious and it was clearly torturous for him to say anything, to allow himself or me even a glimpse of what he had just experienced. However, it was apparent that something new and indefinable, a 'primitive agony', was now present in the field between us. This feeling became a recurrent theme for approximately the next year, and the silent gaps occasionally extended to ten or fifteen minutes. I felt *frozen out* just as he was *frozen in*. The absence of any psychic articulation sometimes felt like a deathlike void. Occasionally, my mind swirled in near panic, and I would feel nauseated during and after the sessions. This embodied interplay evoked no interpretations in my mind, and I knew that I must refrain from inventing something clever to more comfortably coat the experience.

However, repetition is rarely pure. On one occasion, he glared at me angrily and accusingly, suddenly red-faced with emotion, and burst out with the words, 'I just can't figure out what you want!' The emotion quickly subsided, as if it had never happened. When I enquired further, he said he had no idea where that originated, and he did not remember his words. This experience was a clear example of that which Laplanche calls *transfert en creu*, to the empty pot, or the transference to the enigma to the analyst. This transference is frequently a harbinger of the opening of new dimensions. It likely reflects the question that Lacan noted as universal in human development: 'What does the other want?' (Johnston 2013: 28). It felt like the emergence in the transference of the question that had always mystified him. The past was becoming present in articulated form.

Finally, Ralph entered the room one day and lapsed into a painful silence that was different from before. For the first time, he felt fully there, engaged with me in the room. With a piercing, angry look, he burst out loudly, 'What the hell are we doing here?!' With no hesitation, the words emerged from my mouth, 'Fuck if I know!'

I was totally startled by my words, as was he. We gazed at each other as if fixed in a momentary trance, and the room filled with anxiety and dread. Time seemed strangely suspended. Suddenly he flushed and I flushed, and we both broke down in peals of deep belly laughter.

We had both reddened in our mutual gaze, signalling shame in its full embodiment, but it was not the shame of denigration or humiliation. Shame can freeze or paralyse a person in time, or on the other hand violently disrupt the frozen uniformity of things and open the temporal flow. Levinas connects shame, ethics, and temporality as experiences 'in which the Other does not block my future but rather opens it up' (Guenther 2011: 25). That is, abrupt swerves of presence and absence may provoke shame, but that shame is often a marker of sudden shifts, in experience rather than a merely negative symptom.

Bergson (2012) said that laughter removes the crust of the mechanical from the living. It opens the flow of time. André Green (2009: 17) called such moments of breakthrough of temporality, *le temps éclaté*, or 'exploded time'. In this view, linear structures and strictures of time are 'exploded', and a more 'open ensemble' of psychic life emerges (Ibid.). In Ralph's case, he was emerging from the frozenness of a past that had entered the present through the transference. Rather than endless repetitions, our experiences together had become his and our 'history' through enactments of the previously unrepresentable (Marucco 2007). The transference had slowly formed a new inner 'base' for him, enabling the dramatic eruption to occur without danger of complete disintegration. He became more present in our relationship, as did I. There were fewer undertones of pleasing, and this attitude began to generalise to other dimensions of his life. His life felt more inwardly and outwardly populated. When I mentioned the shift in his demeanour and behaviour, he said that it was as if a big weight was off his back.

Events began to progress. He met a new woman, and their relationship appeared to be functioning well. He was much less inhibited and frozen in this relationship.

Approximately six months after the episode of laughter, he had the following series of dreams

> It is a surreal landscape, like a jungle area in Central America where I've done volunteer work. A young woman is walking along a trail on the side of a hill ahead of me, but suddenly disappears off to the side, like into a *cave* or something. I am alarmed, and plunge ahead, wanting to find her and help her. There is a wall of dry board covering the opening of the cave . . . I break through it, looking for the woman. There are some *claw marks* on the wall and I know there is an invisible *monster* in there. I feel absolutely terrified!

He stopped for a moment in his narration, appearing very apprehensive, and mentioned the following association: the young woman reminded him of the photos of his mother as a beautiful, lively young woman. Slowly and seriously, he discussed his lifelong feelings of loss and loneliness and his ongoing terrors as a child – feelings that we had revisited many times. He had never been able to identify what terrified him. The fear had always seemed unknowable, and the more he had tried to contemplate it, the more mentally paralysed he had felt.

He continued the dream narrative

> Then I see some balls – large, blown up plastic balls like children play with – and I think that I can distract the monster. I don't see it but it's there. It seems like I can do it: distract it [with a ball]. Then it seems like the monster is different, like it has become *playful*. The whole mood becomes different.

He said excitedly that it was the first time that he had been able to hold his own and act in the face of such terror, whether in dreams or in everyday life.

These dreams remain remarkable to me. The jungle area is reminiscent of Central America, where he had travelled for quasi-philanthropic reasons. It had likely been his means of leaving the family environment and nourishing a spark of personal adventure. The jungle is nature, or whatever we project as nature. He followed the woman as he has done his entire life, but developed a more active concern when she disappeared. He crashed through a barrier – which is unusual, as he is not an action-oriented person – and found himself in a cave with the monster. This action was the revelation of the way his search for permanence and uniformity had kept him frozen in time. It had had a monstrous effect on his life.

That is, the uniformity and unity that he had been pursuing all his life in the form of the young woman in the photographs was in fact a type of trap. By serving the monster, he had sought *an atemporal Something* that would enable him to escape the flow of time and differentiation. The brief, intense affair that had ended abruptly just before he entered analysis appeared to offer that possibility. That loss felt as if he were losing access to the source of life itself, and he was shattered. Such seductive illusions, like a Return to the Garden, dominate much of human existence and frequently arrest us in a type of prolonged childhood. The loss of these illusions creates shock and disillusionment and, occasionally, a sense of subjective destitution. Such experiences can result in major shifts in the temporal dimensions of the soul (Giegerich 2008; Žižek 2014: 113).

Claw marks comprise the only articulated presence in the cave, and conveys figuration at its most elemental. As in Palaeolithic cave art, the most precious human creations are symbols. Symbolism is the emblem of human participation in time. For Ralph, it signified entry into time at an entirely new level, a transformative experience that affected him at the core (Cassorla 2013).

Playing with the monster was his means of surmounting the terrifying encounter. The capacity for play and creative imagination is frequently crucial in surviving and transmuting the dark 'Things' that erupt in our lives during disturbing transitions (Bakhtin 1981 and 1984; Cassorla 2013; Hinton 2007; Huizinga 1955; Kristeva 1989; Parsons 1999). Inchoate 'primitive agonies' can thus be translated into representation and memory, and an 'illumination' of the original traumatic scene can emerge, fostering a *re-creation of Time* (Scarfone 2006; Winnicott 1963; Stein 2008; Marion 2012). Ralph had been frozen in a state of iteration and sterile sameness, but the dialogical movement of temporality had been restored by the potent emergence of alteration.

The remaining eighteen months of analysis continued in this vein, and more complex and articulate memories emerged. He made two extensive visits to his old hometown, and felt revitalized when he found classmates that he had forgotten. Now, when he visited his sister, I could feel the aliveness and visualize their interactions. Ralph's past was becoming present!

His sense of humour continued to evolve, and his general manner was less pleasing and more real. I typically hear from Ralph once a year. At the one-year anniversary of termination, he sent me a card picturing a Hindu fakir lying on a bed of nails; the fakir had a playful look and was almost laughing. This image appears to be a good summary of the best that one might achieve in analysis. Ralph has remarried and appears to be happy.

Freud's *Nachträglichkeit*: attributing the past with new meanings

What contemporary psychoanalytic perspectives best contribute to understanding the psychological life of someone like Ralph, and for our patients and us? In my experience, most Jungians still tend to think of Freud as extremely reductionistic, and deeply flawed in his understanding of the unfolding of the human personality. It is true that Freud had intended to create an analytic theory with solid foundational principles, principles that could be applied with a certainty that fit the science of his times; but, almost despite himself, he also laid the groundwork for later understandings that undermined his own intent. Based on some of his early writings, post-Freudian theory describes some very important ways of viewing the psyche.

Many of these new developments in psychoanalytic thought circulate around the nature of temporality, especially memory and remembering. From its foundations, an unresolved antinomy regarding time has plagued psychoanalysis, reflecting the two different sides of Freud's own thinking (Johnston 2005: 21; Botella 2014). On the one hand, there is Freud's better-known linear, developmental model in which a fixed and identifiable past underlies the present and future in predictable ways, and another model demonstrating that the past is endlessly shaped and reshaped by future experiences, and that memory is a constant process of re-creation. This is dramatically visible in the case of Ralph, and lies at the heart of much contemporary theory, especially the French School.

Whatever his conscious intentions, Freud was an honest and keen observer. In 1895 (348–59), he noticed the uncertain quality of time and memory when a young woman, Emma, who had a crippling fear of entering shops alone, remembered a disturbing event that had occurred just after she entered puberty. While shopping, she had thought that two of the shop assistants were laughing at her clothes, and she left the shop in a panic. Subsequently she recalled that she had felt sexually excited by one of them. Later, she remembered an incident at age eight when an old shopkeeper had grabbed her genitals through her clothes. However, notably, that original incident had not resulted in a fear of entering shops. Freud reasoned that the second event, which had occurred when Emma's sexual development was emerging, had enabled the first event to be retrospectively understood as a sexual assault. He saw that realisation as the precipitant of the current symptoms.

Freud thereby discovered that the remembering of psychic life could be decisively influenced by an event that never occurred, a lost object that was never possessed, or a primal scene whose power was constituted retroactively (Boothby 2001: 182). That is, he recognized that earlier experiences are subjected to revisions and alterations, and therefore, there is not a fixed past or an immutable time base. *Nachträglichkeit* was the term coined by Freud for these rearrangements of memory that confer the past with new meanings. The term was later translated as 'afterwardness' in English and 'après coup' in French. The nuances of the German word *Nachträglichkeit* are difficult to translate. It means both going backwards and adding something, or a process of being 'drawn after' like pulling a wagon, with a hint of reluctance or even truculence. In psychoanalysis, it has to do with how later experiences may, in effect, retranscribe earlier ones. Such a view highlights remembering as an activity, not merely a recall of an identical datum (Loewald 1972: 408).

Freud, as a scientist of his times, was always searching for foundational certainty in his theories. *Nachträglichkeit* seemed to undermine his quest for unchanging engrams of memory underlying, and therefore finally explaining, psychological experiences. He never provided a promised volume on memory, although he mentioned *Nachträglichkeit* from time to time, notably in a footnote in the case of the Wolf Man.

It was left to the next generation of analysts to further develop the concept of *Nachträglichkeit.* Lacan was the first to revise Freud's theory of memory, asserting that human experience always contains a remainder – a kernel of memory that defies signification. He named this enigmatic element 'the Thing' – 'das Ding', a potent *something* at the core of experience that foments the process of *Nachträglichkeit* (Boothby 2001: 162). This added a new dimension to Freud's concept by pointing to another and more terrifying significance of the old shopkeeper's desire that was beyond mere sexuality for the young girl's psyche. Because of her immaturity, the man's demand had defied representation in her memory. This was due to the gap between her perception of his desire and her capacity to represent this desire. At her level of development, his desire was an unrepresentable enigma. Such a gap in representation creates *a remainder that cannot be symbolized*, a potent 'Thing' that may feel like 'finding a hieroglyph in the desert' (Lacan 1977: 194). For Lacan, it referred to an unknowable background of experience that lies at the core of the subject, a subject who is always asking the unanswerable question: 'What does the other want of me?' (Ibid.: 198–209). This lends a profound complexity to Freud's discussion, which emphasized only the development of sexual desire. It becomes not only a matter of a simple difference in sexual maturity between the child and the adolescent, but of something *untranslatable* and therefore more disturbing. From this point of view such memories have a core that can provoke endless translations, but can never be finally located in time. For Lacan, this aspect of *Nachträglichkeit* assumed a primary significance.[5]

Laplanche later emphasised the intersubjective element of enigma, theorising that some memories are untranslatable because they are based on messages from parents to children that are communicated in ways and on levels – particularly sexual levels – that are unconscious. The adults transmit messages they themselves do not understand (Target 2007). Such untranslatable messages are managed by means of what Freud called 'primal repression', in contrast to the everyday sort of 'secondary repression' that assumes a capacity for representation (Laplanche and Pontalis 1973: 333–34; Hinton 2009). Laplanche emphasised the helplessness of children regarding such messages (*hilflösigkeit*). Children are dependent on others for survival; they lack defences and do not understand all the levels of the adult communications directed toward them. The core of the primally repressed can never be fully articulated, and serves a role akin to drive (*trieb*). That is, when such untranslatable memories of 'otherness' re-emerge later in life, they often disrupt repetition and sameness, violently thrusting the individual toward an unknown future requiring new understandings that in turn, create new memories.

Ralph's story describes such a process. Like the claw mark he discovered in the cave, such memories are impossible to finally translate, but provoke a process of remembering that is central to subjectivity, and is continuous throughout our lives (Laplanche 1999: 117–33). It is both a kind of freedom and limitation. There is no final 'knowing' except for lifelong translations and retranslations. This process can be lived playfully or tragically.

The neuroscience of memory: 'the remembered present', time travelling and *autonoetic* consciousness

In conclusion, I want to relate the remarkable coincidence of recent discoveries in the neuroscience of memory and contemporary psychoanalytic theory. Such studies similarly describe a process that is never completed and is subject to revision each time a 'memory' is elicited (Civitarese 2010; Fernyhough 2012). Edelman calls memory the 'remembered present' (Edelman 2004: 99). That is, past ideational traces can never be recovered in an unmodified form within the constraints of the present. These 'traces' are not like stable atomic units; they are moments caught in an ever-mobile Now (Johnston 2005: 315).

We frequently fear that the 'ghosts' of the past will re-appear in the future, and the future might then encompass that which will have been remembered. In the future, we might remember and, thereby, produce a past. Such perspectives can seem a bit overwhelming and counter-intuitive at first glance, but they have strong grounding in neuroscience and metapsychology.

Endel Tulving of the University of Toronto, with others, has produced fascinating work in the area of memory and that which he half-humorously calls 'time travel'. His research indicates that human beings possess a specific form of memory involving the capacity to be consciously aware of one's continuing temporal existence. This type of memory is called *episodic* memory, in turn related to a form of consciousness known as *autonoetic consciousness.* No other animals seem to possess these characteristics (Szpunar and Tulving 2011: 84). They are fundamentally different from a more general type of conscious awareness known as *semantic consciousness* (Ekstrom 2014: 37–38). 'The essential difference [between semantic consciousness and autonoetic consciousness] . . . is between knowing that something *is* such and such, or [*factually*] occurs so and so . . . [versus] remembering that *one had an . . . experience* (witnessed, felt or thought something) in a particular way at a particular time, within a particular human context' (Tulving 2005: 16; Ekstrom 2014: 34–36).

Episodic memory, the capacity for 'time travel', originates in semantic memory and adds the ability to travel in *time* to an already existing ability to mentally manoeuvre in *space*; it provides the capacity to 'travel' into the past and, more importantly, into the *future*. Thereby, *humans became able to create a world to fit them*, rather than live in one into which they had to fit (Tulving 2005: 17; Suddendorf, Addis and Corballis 2009).

Many neuropsychological and developmental studies support the division between semantic time and a later-emerging episodic time, providing detailed evidence for this theory (Hsieh *et al.* 2014). One notable example is the case of a 30-year-old man, K.C., who suffered a brain injury that destroyed his capacity for episodic memory and, therefore, for travel into the past or future. 'However hard he tries, and however powerfully he is prompted, he cannot bring into his conscious awareness a single event, happening or situation that he witnessed or in which he participated. Nor is he capable of remembering anything ever having happened in the house where he has lived for [many] years. He knows that he owned a black Honda, but does not remember a single trip he ever took in it' (Tulving 2005: 24). It is particularly striking that K.C. has no thoughts regarding his future, which highlights his lack of autonoetic consciousness. He possesses many intact cognitive capacities regarding facts; however, he does not remember any episode from his life nor project himself into the future. When asked to do either, he states that his mind is 'blank, but a *blankness* only with regard to memories or prospective futures' (Szpunar and Tulving 2011: 84).

Episodic memory is also largely missing in children under the age of four (Ibid.: 29–30). They do not remember personal contexts well, and the ability to remember interpersonal events comes after the emergence of knowledge regarding facts. Younger children lack the same types of episodic memory and autonoetic consciousness that appear to be absent in animals (Szpunar and Tulving 2011: 31; Yim, Dennis and Sloutsky 2013; Ekstrom 2014: 42–44). Before around 18 months, children will pay attention to their mirror image but do not realise it is their body, and will not use the mirror as a tool to think about themselves. However, passing the mirror test does not mean that they are aware of themselves as individuals with a past and a future. Episodic memory matures gradually, and it is difficult to pin down an exact age when it becomes fully functional.

However, the general finding is that it is around four years of age (Tulving 2005: 31; Yim *et al.* 2013).

The drawer test helps clarify the time of emergence of episodic memory. In one example three- and five-year-olds are told, shown, or given hints of the contents of a drawer. Within a short time, they are asked what is in the drawer, and then how they knew what they knew. Most of the children in both age groups recalled the facts of what the contents were, but there was a large age difference regarding how they believed they had come by that knowledge. Most significantly, the five-year-olds remembered that someone had just told them what was in the drawer. In contrast, the three-year-olds mostly did not remember being told what was in the drawer. That is, the five-year-olds recalled an event in the past in an interpersonal context, the essence of episodic memory, while the younger children recalled only the facts without a sense of past or context. It appears that young children cannot, or do not, represent their knowledge as deriving from a particular time in the personal past, but rather as detached, impersonal knowledge. For them, 'There are [just] crayons in the drawer' (Tulving 2005: 32).

Such findings have been replicated in a number of ways. The growth of episodic memory apparently continues to increase with age, even into adulthood (Yim *et al.* 2013). 'The developmental findings not only emphasize the role that a fully developed human brain plays in episodic memory; they also underscore the fact that episodic memory is not necessary in a world in which one's needs can be and are satisfied without remembering autonoetically what happened in the past' (Ibid.: 35). The case of K.C. provides a similar perspective.

Recent MRI studies demonstrate that specific brain areas are connected to the experience of past and future and that these areas are located together. Contemplating the prospective future and actively remembering the past result in activation of identical areas of the brain (Nyberg *et al.* 2010; Szpunar and Tulving 2011). These findings reinforce concepts that I discussed earlier – *Nachträglichkeit* and the past as the 'remembered present'. Earlier, I pointed out the intimate and non-linear connection of past, present and future. Derrida describes temporality as being more like a syncopated ensemble, or in Serres' words, resembling a folded and crumpled handkerchief (Serres 1997: 44; Pirovolakis 2010: 43–82).

Mental time travel is a source of cultural evolution, as well as the origin of tragic aspects of the human condition. Human beings worry regarding the future, and we struggle to discern the threads of patterns that were or might be. This worry has clear adaptive aspects and is the basis of culture and ethics. Ethics involves reflection upon the effects we might have had on others in the past and that we might have in the future. We have fears regarding future dangers, and we dwell on the past. Could certain troublesome others become dangerous to us or to our group? Should we eliminate them from the utopias we plan? Will the speed of technology entail a transcendence of Human Time (Stiegler 1998: 15; Fong 2013)?

At the core of all this is a view of *analysis as a complex temporal process*, a future that may alter the past, a past that may become the future, simultaneously existing along with an evanescent present as the pulsating intersection of past and future. This aspect of time is syncopated, multifaceted and non-linear. Shocks, disruptions, melancholy and enigma are at the heart of human experience and consciousness. However, embracing the perturbations inherent in temporality can open an element of freedom, and a sense of creative play in meeting the strange and tormenting, the delightful and ever-new vicissitudes of time (Parsons 1999). This was what my patient Ralph discovered, to his great gain.

Notes

1 This chapter has been previously published in the *Journal of Analytical Psychology* 60(3): 353–70 (2015).
2 Laplanche speculates that what moves the 'Dichter' (poet) to 'Dicht' (poetise) is the desire to address another who is out of reach, or others 'scattered in the future' (Laplanche 1999: 223–24).
3 Many authors use *Zürückphantasieren* to describe Jung's theory of transforming memory. Strangely, this term does not appear in the *Collected Works*, but it does seem to convey Jung's perspective.
4 This case was discussed from a different perspective in *The Journal of Analytical Psychology* 54: 637–57 (2009).
5 The concept of *Nachträglichkeit* strongly influenced Jacques Derrida, and was of importance in the formation of his basic concept of 'différance'. It was the impossibility of a foundational ground in memory and experience that caught Derrida's eye. Through him, Freud's early ideas strongly influenced post-modern philosophy (Derrida 1998; Hodge 2007: 8–9).

References

Bakhtin, M. M. (1981). *The Dialogic Imagination*, Austin, University of Texas Press
——. (1984). *Rabelais and his World*, Bloomington, Indiana University Press
Bergson, H. (2012). *Laughter: An Essay on the Meaning of the Comic*, CreateSpace Independent Publishing Platform
Boothby, R. (2001). *Freud as Philosopher: Metapsychology after Lacan*, New York, Routledge
Botella, C. (2014). On remembering: The notion of memory without recollection. *International Journal of Psychoanalysis* 95(5): 911–36
Brooks, R. M. (2011). Un-thought out metaphysics in analytical psychology: A critique of Jung's epistemological basis for psychic reality. *Journal of Analytical Psychology* 56: 492–513
Campbell, J. (1983). *The Way of the Animal Powers, Volume 1*, New York, Harper and Row
Caruth, C. (1996). *Unclaimed Experience: Trauma, Narrative and History*, Baltimore, Johns Hopkins University Press
Cassorla, R. M. S. (2013). In search of symbolization: the analyst's task of dreaming, in H. B. Levine (ed.), *Unrepresented States and the Construction of Meaning*, London, Karnac Books, 202–19
Civitarese, G. (2010). Nachträglichkeit, in *The Intimate Room*, New York, Routledge
Derrida, J. (1998). *Archive Fever: A Freudian Impression*, Chicago and London, University of Chicago Press
Edelman, G. (2004). *Wider than the Sky: The Phenomenal Gift of Consciousness*, New Haven, Yale University Press

Ekstrom, S. R. (2014). *Memory and Healing: Neurocognitive and Psychodynamic Perspectives on How Patients and Psychotherapists Remember*, London, Karnac

Faulkner, W. (1951). *Requiem for a Nun*, New York, Random House

Fernyhough, C. (2012). *Pieces of Eight: The New Science of Memory*, London, Profile Books

Fong, B. Y. (2013). Death drive sublimation: A psychoanalytic perspective on technological development. *Psychoanalysis, Culture and Society* 18(4): 352–67

Franz, M. -L. (1974). *Number and Time: Reflections Leading toward a Unification of Depth Psychology and Physics*, Evanston, Northwestern University Press

Freud, S. (1950/1895). *Project for a Scientific Psychology*, SE Vol. I, London, Hogarth

Giegerich, W. (2001). *The Soul's Logical Life*, Frankfurt, Peter Lang

——. (2008). *Soul-Violence: Collected English Papers. Vol. III*, New Orleans, Spring Publications

Green, A. (2009). From the ignorance of time to the murder of time. From the murder of time to the misrecognition of time in psychoanalysis, in L. T. Fiorini and J. Canestri (eds), *The Experience of Time: Psychoanalytic Perspectives*, London, Karnac

Guenther, L. (2011). Shame and the temporality of social life. *Continental Philosophy Review* 44: 23–39

Hinton, A. L. (2014a). Book review essay of Time and Timelessness: Temporality in the theory of Carl Jung, by Angeliki Yiassemides. *Journal of Analytical Psychology* 59(3): 437–47

——. (2014b). Justice and time at the Khmer Rouge Tribunal: In memory of Vann Nath, painter and S-21 survivor. *Genocide Studies and Prevention: An International Journal* 8(2): art. 5

Hinton, L. (2007). Black holes, uncanny spaces and radical shifts in awareness. *Journal of Analytical Psychology* 52: 433–47

——. (2009). The enigmatic signifier and the decentred subject. *Journal of Analytical Psychology* 54: 637–57

Hinton, L., III, Hinton, L., IV, Hinton, D. and Hinton, A. L. (2011). *Unus Mundus:* Transcendent truth or comforting fiction? Overwhelm and search for meaning in a fragmented world. *Journal of Analytical Psychology* 56: 375–96

Hodge, J. (2007). *Derrida on Time*, New York, Routledge

Hoy, D. C. (2012). *The Time of Our Lives: A Critical History of Temporality*, Boston, MIT Press

Hsieh, L.-T., Gruber, M. J., Jenkins, L. J., and Ranganath, C. (2014). Hippocampal activity patterns carry information about objects in temporal context. *Neuron* 81: 1165–78

Huizinga, J. (1955). *Homo Ludens: A Study of the Play-Element in Culture*, Boston, Beacon Press

Johnston, A. (2005). *Time Driven: Metapsychology and the Splitting of the Drive*, Evanston, Northwestern University Press

——. (2013). Jacques Lacan. In *Stanford Encyclopaedia of Philosophy Archive. Summer 2014 Edition*, ed. E. N. Zalta. Retrieved from http://plato.stanford.edu/archives/sum2014/entries/lacan/

Jung, C. G. (1930). *Some Aspects of Modern Psychotherapy*, CW Vol. 16

Kristeva, J. (1989). *The Black Sun*, New York, Columbia University Press

Lacan, J. (1977). *Écrits. A Selection*, London, Tavistock/Routledge

Laplanche, J. (1999). *Essays on Otherness*, New York, Routledge

Laplanche, J. and Pontalis, J. -B. (1973). *The Language of Psychoanalysis*, New York, W.W. Norton

Leroi-Gourhan, A. (1967). *Treasures of Prehistoric Art*, New York, Harry Abrams

Lin, Y. (2013). *The Intersubjectivity of Time: Levinas and Infinite Responsibility*, Pittsburgh, Duquesne University Press

Loewald, H. W. (1972). The experience of time. *Psychoanalytic Study of the Child* 27: 401–10

Marion, P. (2012). Some reflections on the unique time of Nachträglickeit in theory and clinical practice. *International Journal of Psychoanalysis* 3: 317–40

Marucco, N.C. (2007). Between memory and destiny: repetition. *International Journal of Psychoanalysis* 88: 309–28

Merleau-Ponty, M. (2012 [1945]). *Phenomenology of Perception*, trans. D. Landes, New York, Routledge

Nyberg, L., Kim, A. S. N., Habib, R., Levine, B. and Tulving, E. (2010). Consciousness of subjective time in the brain. *Proceedings of the National Academy of Sciences* 107(51): 22356–9

Parsons, M. (1999). The logic of play in psychoanalysis. *The International Journal of Psychoanalysis* 80: 871–84

Pirovolakis, E. (2010). *Reading Derrida and Ricoeur*. Albany: SUNY Press

Scarfone, D. (2006). A matter of time: actual time and the production of the past. *Psychoanalytic Quarterly* LXXV: 807–34

——. (2011). Repetition: between presence and meaning. *Canadian Journal of Psychoanalysis* 19: 70–86

Serres, M. (1997). *The Troubadour of Knowledge*. Ann Arbor, University of Michigan Press

Stein, R. (2008). The otherness of sexuality: excess. *Journal of the American Psychoanalytic Association* 56: 43

Stiegler, B. (1998). *Technics and Time, 1: The Fault of Epimetheus*, Stanford, Stanford University Press

Suddendorf, T., Addis, D. R. and Corballis, M. C. (2009). Mental time travel and the shaping of human mind. *Philosophical Transactions of the Royal Society B: Biological Sciences* 364: 1317–24

Szpunar, K. K. and Tulving, E. (2011). Varieties of future experience, in M. Bar (ed.), *Predictions in the Brain: Using our Past to Generate a Future*, New York, Oxford University Press

Target, M. (2007). Is our sexuality our own? A developmental model of sexuality based on early affective mirroring. *British Journal of Psychotherapy* 23(4): 517–30

Tulving, E. (2005). Episodic memory and autonoesis: uniquely human?, in H. S. Terrace and J. Metcalf (eds), *The Missing Link in Cognition*, New York, Oxford University Press

Winnicott, D. W. (1963). Fear of breakdown, in C. Winnicott, R. Shepherd and M. Davis (eds.), *Psychoanalytic Explorations*, Cambridge, Massachusetts, Harvard University Press

Yiassemides, A. (2014). *Time and Timelessness: Temporality in the Theory of C. G. Jung*, London and New York, Routledge

Yim, H., Dennis, S. J. and Sloutsky, V. M. (2013). The development of episodic memory: items, contexts and relations. *Psychological Science* 24(11): 2163–72

Žižek, S. (2006). *The Parallax View*, Cambridge, Massachusetts, MIT Press

——. (2014). *Event: Philosophy in Transit*, London, Penguin Books

4

LIVING *KAIROS*

The time factor in waking dreams

Laner Cassar

Waking dreams are the *via regia* to unconscious material. Time is a very important element in waking dreams, since it helps to define them: the patient pauses real life and enters an atemporal mythical one, as to be able to re-create oneself. This chapter explores the notion of time in waking dreams as described by Carl Jung in his technique of *active imagination*, and in Robert Desoille's *Directed Waking Dream* method. In my exploration of waking dreams, I will focus on the subjective experience of time while a person is engaged in them, and on key therapeutic issues related to the time factor.

A waking dream is different from a daydream. A daydream is usually a passive imaginative activity used as a wish-fulfilment or as an escape (for example from a monotonous situation). Contrastingly, a waking dream (or a waking fantasy) carries an element of choice since one actively chooses to start to fantasize and to engage with the images. It has a purposeful aim since one chooses to imagine and to allow any image to come up, for instance when thinking about a particular fear or a problem.

When speaking of waking dreams one tends to think more of the inner psychic space material explored, and less about the time dimension – their temporal pulsations, particular rhythms, and paradoxes of their duration. As existentialism teaches us, time is primal in human existence, and its quality is constant motion; however, our error in the modern day has been to think of ourselves primarily in terms of space, as though we were objects that could be located like substances at this or that place. As a result, we have lost our genuine existential relation to ourselves, and also with others around us. The French philosopher Henri Bergson (1859–1941) explains that with this overemphasis on spatial thinking 'the moments when we grasp ourselves are rare, and consequently we are seldom free' (Bergson 1888/1960: 231–32). Waking dreams seem to offer a unique moment, a *kairos* in the temporal process, moments in which something unique can happen or be learnt or accomplished.

I will be referring to the waking dreams as described by Carl Jung (1875–1960), namely his psychotherapeutic technique of *active imagination* as well as those described by his contemporary French therapist Robert Desoille (1890–1966) known as the *Rêve Éveillé Dirigé* (RED)/*Directed Waking Dream* method. These two waking dream approaches are similar yet distinct. For Jung, active imagination is a form of spontaneous wakeful dreaming, done alone and with eyes open; while for Desoille RED is a form of dreaming while awake in the presence of a therapist, guided and directed by the therapist, usually done with eyes closed. Carl Jung developed his active imagination method in 1912, while Robert Desoille realised his method independently of Jung in 1921. Desoille corresponded with Jung in March, 1938 and in the late 1930s integrated some of Jung's theory in his work (Cassar 2014). Both were pioneers in the use of imagination in psychotherapy at a time dominated by rationalism and empiricism.

Defining waking dreams according to C. G. Jung and Robert Desoille

Jung disagreed with Freud on the notion of *fantasy thinking* as a symbolic disguise of instincts, and on Freud's often-quoted statement that 'a happy person never phantasises, only an unsatisfied one . . . and every single phantasy is the fulfilment of a wish, a correction of unsatisfying reality' (Freud 1908: 146). In *Psychological Types* Jung differentiated two types of *fantasy thinking* or *mythic thinking* (he used the two terms interchangeably). He spoke of 'passive fantasy' (fantastic imagination) and 'active fantasy' (true imagination) (Jung 1921/1971: para. 712). Passive fantasy is 'a subjective figment of the mind' while active fantasy is 'an image-making, form-creative activity' (Jung 1967: para. 207). For Jung

> [Fantasy] remains the creative matrix of everything that has made progress possible for humanity. Fantasy has its own irreducible value, for it is a psychic function that has its roots in the conscious and the unconscious alike, in the individual as much as in the collective.
>
> (Jung 1921/1971: para. 86)

For Jung waking dreams or fantasies are 'evoked by an attitude directed to the perception of unconscious contents, as a result of which the libido immediately invests all of the elements emerging from the unconscious' whereas in the case of 'passive fantasy' this does not occur (Jung 1921/1971: para. 712).

Jung referred to the use of waking dreams for therapeutic purposes as *active imagination*. In *The Psychological Aspects of the Kore*, he defined active imagination as 'a method of introspection for observing the stream of interior images' (Jung 1951/1968a: para. 319). He also gave some guidelines to what can be a starting point of active imagination: 'Take the unconscious in one of its handiest forms, say a spontaneous fantasy, a dream, an irrational mood, an affect, or something of the kind, and operate with it. Give it your special attention, concentrate on it' (Jung 1955/1970: para. 749).

Once this process is underway, Jung explained, by 'concentrating on the image it begins to stir, the image becomes enriched by details, it moves and develops' (Jung 1935/1976: para. 398). This eventually leads to a whole *oneiric drama* in which the subject will also converse with encountered images. In the classic Jungian sense, active imagination is done on one's own, although it is nowadays included in Jungian analysis.

For the French psychotherapist Robert Desoille who also used the imagination in his therapeutic work, the waking dream is 'an intermediary stage between the waking state and the dreaming state, between the physiological and the psychical' (Desoille 1961: 10). He compared this state to Freud's early use of the 'pressure technique' (Ibid.: 21). For Desoille, the waking dream provides us with an 'intimate language', as described by the French philosopher Georges Politzer, about the patient's affective interior life, including his emotional deficits, psychic conflicts and desires. For Desoille the best way to help the patient to have a waking dream is while laying on a couch or an arm-chair with his eyes closed, in a dimly lit environment thus facilitating regression (Desoille 1938: 46). Desoille would give a word stimulus to the patient, such as a descent into the sea: 'Imagine you are at a seashore, a rocky coast where the water is very deep, wear a diving suit and descend far into the water reporting all that you visualise and experience as you explore the sea.' Other stimuli could be: climbing a mountain, an encounter with a dragon, etc. where the patient has to let his or her imagination work on the stimulus and create an inner oneiric drama in which he or she is present while voicing out-loud to the therapist whatever he or she is witnessing, feeling, and experiencing. Desoille would encourage the patient to explore the heights and the depths of the inner space, since he believed that going upwards would lead to a sublimated feeling.

In the waking dream the patient would pause from his or her real life and enter an atemporal mythical world, echoing Coleridge's 'willing suspension of disbelief' while remaining conscious all along. In waking dreams the 'here' is elsewhere and places are telescoped, losing both their geographic and geometric context. More than anything else, *the time of the image* blends past-present-future into *pure duration*. The idea is not to remain in a timeless world to escape the real world, but rather to have an experience in the space of the time-controlled therapeutic environment.

Waking dreams and the unconscious

A timeless quality

The waking dream taps into the unconscious, becoming its *via regia*. Through the waking dream the patient can get in touch with his or her complexes and deal with them. The oneiric drama is made up of fantastical figures and theriomorphic beings, which alternate with persons from one's current or past life. Jung in *The Red Book* – his personal journal of his encounters and dialogues with his unconscious images and visions which he also portrayed in drawings – explained that he had to engage with his inner fantasies: 'I must catch up with a piece of Middle Ages – within myself' (Jung 2009: 330). Through the images from the collective unconscious the timelessness of the unconscious circle replaces linear time; beginnings are endings and endings are beginnings; one can go backward as easily as forward, and that which has been lost and forgotten can be retrieved. Past, present and future can come together, forming links and meanings, and producing intense experiences for the subject that may initiate qualitative psychic changes and make a significant difference in one's internal life. The symbols met in the waking dream do not only represent repressed unconscious material from one's personal life, they also indicate *directions for the future.*

Both Jung and Desoille were forward looking (unlike Freud) and they were more interested in the personality development of healthy individuals. They also saw their methods as a form of teaching of the art of living, open to all ages. According to Jung, 'time comes apart a little in the unconscious, that is, that the unconscious always remains beside the passing of time, and perceives things that do not yet exist' (Jung 2008: 9–10). As Yiassemides explained, 'For Jung the past is vital, but of equal importance is the future . . . Jung clearly stated his belief regarding the temporal directionality of the psyche: "Life is a flux, a flowing into the future, and not a stoppage of backwash" [Jung 1951/1968b: para. 278]. There is a dynamic, purposeful movement forward. Retrospection is necessary so that the psyche can stay rooted in the past and thus travel forward toward its telos of self-realisation' (Yiassemides 2014/2016: 22).

Thus, the missed opportunities of the long-ago past may become transformed through the waking dream into a *potential field of dreams* where it becomes possible to mentally time travel, to redeem old disappointments into a better future. Through the waking dream a person leaps from his anchored place and fate to live out his one and only life, and refigure the frustrations and disappointments of the past into a new world of possibilities offered by the future.

For Desoille, movement in space spelled metamorphosis. He did not discover the imagination but the 'verb' of imagination, i.e., the fact that when one moves in an imaginative space one would simultaneously experience emotional and physiological changes. The Uruguayan psychiatrist Prof. Mario Berta, a disciple of Desoille, stated in his unpublished manuscript *Les images qui guerrisent* (*The images that Heal*) that personal maturity occurs in three directions: 'Forward (anticipation), above (ascension) and sideways (polarisation). The existential evolution is a rather complex process which implies the simultaneous anticipation of the future with precise anticipations, of rising towards higher levels of consciousness and a polarisation of opposing characteristics of the world in which man lives' (Berta 2000: 12, my translation).[1]

For Desoille, the ascensional movement is very crucial and its frequency and quality manifests the degree to which the patient has been freed from his/her neurotic problem that has brought him/her to therapy. Desoille grants a special privilege to ascents since the vertical dimension reveals the temporality, authenticity and historicity of one's existence. In the waking dream one abandons the anthropological understanding which views man as man in his human world, and accedes to an ontological reflection which concerns the mode of his existing being as presence in the world.

Rate of time's flow

The most common modification in the temporal experience in waking dreams has to do with changes in the rate of time's flow. In waking dreams, time may be experienced to run faster or slower than it normally does. This depends on the ability of the patient to enter the waking dream, as well as on the positive or negative contents involved in it. The prevalent experience is that in what objectively (i.e. as measured by the clock) is a short period of time much can be experienced; even though so much seems to have happened in the waking dream, in 'reality' very little time has passed. What we have here may be conceived of in terms of two frames of reference: In the ordinary frame of time hardly any time has passed, whereas in the other, non-ordinary frame, one seems to have traversed a very long duration. In other words, it is as if in brief moments a person can accommodate what is experienced as long stretches of time. Theoretically, in the waking dream, experiences of timelessness associated with unbounded duration may be experienced. In this sense, the patient will find him/herself to be outside of the dominion of time.

Experimental work on the relationship between perceived and measured time began under the influence of the psycho-physical notions of Gustav Fechner. This topic was also tackled by various literary writers. For instance, similar observations are found in the fantasy tales of C. S. Lewis' *The Lion, the Witch and the Wardrobe* (Lewis 1950), in which the author recounts the adventures of children who passed through a wardrobe and entered magical lands. The children's adventures in these other lands lasted hours, days, perhaps even longer; yet, as far as the ordinary world was concerned, hardly any time elapsed at all. Thus, when the children returned to the wardrobe and into their house they realised that the adult members thought that they were absent for only a moment or two. Such experiences are also found in the taking of psychedelic drugs such as 'mescaline' and 'ayahuascha'. Interestingly, Desoille experimented with 'peyotl' and with 'mescaline', whereas Jung was against the use of drugs as a method of tapping into the unconscious. Jung was even against the use of drugs such as mescaline (which he had learnt about from Dr Prinzhorn) to alter states of consciousness, since with drugs you can manipulate the brain so that it produces so called 'spiritual experiences'. In the addenda section (1906–1950) to the second volume of Jung's letters, in a letter to Christian Stamm (dated 23 April 1949) Jung affirms that there is no need 'to try and loosen up the unconscious' however, he gives an exception when he says 'though an honest drink would none forbid' has been held sacrosanct from time immemorial (Jung 1973: xlv–xlvi).

Working towards *kairos* in waking dreams

The right time to introduce waking dreams

The right moment, or what the Ancient Greeks called *kairos*, of when to start using waking dreams is also a very important issue. For Jung, waking dreams had to be introduced after an analytic process has been performed to ensure that the patient's ego is strong enough to face the unconscious forces. However, later in his career Jung relaxed this isolationist view of active imagination and stated: 'The inner dialogue could just as well take place between patient and analyst' (Jung 1955/1970: para. 186).

Nonetheless, one should be careful so as not to rush the patient into waking dreams when in therapy; also, a therapist needs to be trained and have experienced it himself/herself during therapy. One should first build a relationship and explain the method before using it. Both Desoille and Jung emphasised the need of a quiet atmosphere. For Jung, stillness helps to produce an *abaissement du niveau mental*, i.e. a lowering of ego-consciousness, a required condition to catch one's interior fantasies. Desoille preferred a semi-dark room, the patient on a couch and the therapist helping the patient to relax.

Ideally, an opportunity arises for the patient to enter spontaneously in a waking dream. For instance, if a patient describes his/her current state as if he/she is in a closed room, the therapist can encourage the patient to explore this image, i.e. to describe in detail this room and to make an oneiric drama from it. Jung encouraged a continuous dialogue with encountered figures; Desoille would also encourage moving upwards and downwards along a vertical axis in the imaginative inner psychic space.

Desoille was one of the first therapists who used the notion of brief dynamic therapy even prior to the one pioneered by psychoanalysts A. French and T. M Alexander, known as brief dynamic psychotherapy (Alexander and French 1946). Desoille would see patients not more than once a week and at times he would even give two or three weeks in between each session. His individual session could last from one and a half hours to two hours and he would spread them out between twenty to forty sessions. Nowadays sessions are no longer than an hour with about twenty minutes devoted to the actual waking dream.

In his 1945 book, Desoille explained how some of his patients had been in psychoanalysis for years with very little improvements. After a few sessions Desoille's patients were reporting dramatic feelings of relief and well-being. This would coincide with the patient reaching new height levels in his RED which promoted the patient's independence from the therapist who had served as a guide. Moreover, these images would become spontaneously generated, and his need to be there to induce them diminished. Nonetheless, Desoille at times would describe his method as complementary to psychoanalysis and in certain cases not ideal, as when there was strong resistance to engage in a waking dream.

The right time to challenge undesirable encounters in waking dreams

The client is encouraged to engage with inner imaginal figures and to speak to them since this provides an opportunity to integrate disowned aspects. Jung himself had to move away from 'the spirits of the time' and move to 'the spirit of the depths' through active imagination since:

> The spirit of this time has condemned us to haste. You have no more futurity and no more past if you serve the spirit of this time. We need the life of eternity. We bear the future and the past in the depths. The future is old and the past is young.
>
> (Jung 2009: 253)

When in waking dreams, patients are bound to meet negative figures. Jung met *Salome* and *the Red One* who perturbed Jung. At times these figures are resisted since the client is afraid of them. However, ideally one should learn to face one's fears. Within the oneiric dramas experienced by the patient, affective disturbances are projected in two ways: as deformations of the imaginative corporal ego/surrogate ego which refers to one's representation of oneself in the waking dream itself, or the imaginative corporal ego may remain intact while the disturbing affective components are projected as elements in the other parts of the image (e.g., as threatening figures such as wild animals, monsters, etc.).

In facing anxious situations in the waking dream the therapist encourages the patient to hold strong and not to run away from the fearful situation and to keep on dialoguing with the encountered figures. These moments can afford the possibility of a transcendence of the temporal domain and lead to a moment of *kairos* or working through. *Kairos* is defined here as a passing instant when an opening appears which must be driven through with force if success is to be achieved. *Kairos* implies a right time in the course of events to do certain things that will favour a crucial happening and opportunity which must be immediately recognized and seized upon.[2]

According to von Franz, Jung would never interfere and would leave the patient to find his or her own solution; he would not offer any sort of magic to undo the complex challenges being faced. Von Franz mentioned an episode when Jung described the work with one of his female patients who in her active imagination did not know how to get to the other side of the wall. The patient felt stuck for a long time, until eventually she thought of a hammer to make a hole in the wall. Von Franz argued that had Jung helped her, she would have remained infantile and passive. One has to face one's issues using one's own internal resources (von Franz 1980: 92).

The key attitude is one of trust and recognition of confusion, rather than domination or arrogant assumptions of expertise. In the *Visions Seminars*, when Christiana Morgan reported to Jung an attack by a python, Jung commented: 'one must stand the onslaught of panic, the terrible horrifying moment of attack by the python must be endured' (Jung 1998: 231). By being able to consciously face and tolerate the panic aroused by unconscious figures one begins to thaw the dark depths of one's icy nature, revealing a numinous source of force. Accepting oneself not only requires enormous humility but the overcoming of the temptation to find an easier way out. Von Franz criticised Desoille, Leuner and Happich of employing directed waking dream methodologies to bypass the hard work one is called to do to individuate (von Franz 1980). A person 'would rather devise any trick to help him escape, since nothing matches the torment of one's own way' (Jung 2009: 310a). For Jung if one is to carry out a personal transformation, he/she must let oneself experience that which is scary, if need be, to be bitten by wild animals without running away from them. In doing this one would be exposing oneself to the instinctual impulses of the unconscious without identifying with them or having them flee back to the unconscious.

On the other hand, Desoille's RED seems to be more supportive of the patient to face his or her dark side. Desoille cured patients who had failed to recover via psychoanalysis. In RED, the patient undertakes a gradual regression emboldened by the therapist's sense of security. At times, when the scenario became too threatening for the patient, the patient would try to find a way out so as not to face the menacing force of the images.

Desoille followed Janet's early therapeutic attempts with hysterics, in which he would enter his patient's imagery after they were hypnotized and help them by influencing and changing the basic pathological structure of the experienced fantasy. In fact, Desoille would offer the patients protective objects of '*caractère magique*' (magical character) to face the fearful situation; or ask them who they were or what they wanted from him; or he would ask the patient to imagine the analyst beside him or her.[3] Desoille believed that one had to respond and tackle the imaginary by using his or her own imaginary language (Desoille 1955: 91).

Post-Desoillians remain more facilitative and less directive while employing a form of 'Socratic imagery'. Socratic imagery is essentially Socratic questioning applied in the context of imagery modification, whereby patients are encouraged to create their own mastery/coping imagery, as well as their own self-calming, holding imagery (Smucker 1997). Such coping imagery is thought to offer greater empowerment in moments of anxiety and fear than coping/mastery imagery that comes from the therapist. Other post-Desoillians (Virel and Fretigny 1968; Balzarini 1985) ask the patients to physically ground their imaginative self by becoming more aware of their imaginative surrogate-ego, especially via their feet. Virel and Fretigny (1968) would encourage the patient to move out of the immersive imaginative experience by just watching it as if watching a film. This distancing technique has been recently utilised by trauma therapist Reddemann (2008) when trauma patients are flooded with visual flashbacks of traumatic content. The 'screen technique' and the 'observer technique' are examples of such distancing techniques. Kluft (1989) also describes several 'temporizing techniques' including imaginative ones, to offer the patient respite and temporary asylum to regain mastery and self-efficacy.

Given Jung's own journey, during which he risked losing himself in the unconscious, all therapists should be very careful in how to allow patients to explore their darkest sides. Desoille's RED and the developments of his disciples offer examples how to take such risks in therapy. However, the issue of when the therapist is to encourage the subject to face the dangerous aspects in the waking dream, or when to bring the subject back to safety, is an art in itself. The therapist's effectiveness regarding *kairos* in psychological moments requires that he has done his own personal therapy. This would give the therapist a better sense of when the possibilities for a *kairos* would be greater. He would see when it is necessary to press forward even though this is bound to cause the patient a certain degree of anxiety. Only then can the therapist take a chance and a reasonable risk; this requires a strong level of intuition, flexibility, and creativity.

Charging forward when the time is ripe leads to insight. Desoille spoke of moments when a patient facing an underwater large octopus was encouraged to bring her up to the surface of the water where it changes into the image of the negative parent. Thus, through the waking dream direct insight is also possible, and one does not necessarily have to wait for the next session to find meaning in its content. This is currently more in line with post-Jungian thinking and post-Desoillian approaches. Insight provides a *kairos*, a moment, a chance and an opportunity. Michael Sells describes 'the moment' as 'time out of time within time . . . a universe of promise and peril' (Sells 1996: 100). The Greeks also used the word *kairos* to describe the fullness of a sacred moment-of-truth or revelation and in this sense of time and space it captures the powerful impact of facing one's unconscious.

On vertical time: insight during and after the waking dream

Insight in therapy and through the waking dream seems to introduce a different temporal dimension which we can refer to as a *vertical*, using Bachelard's term of 'vertical time'. Where prosaic time is horizontal and continuous (like Bergson's), the waking dream's time is discontinuous and disruptive, focusing on vertiginous heights and abyssal depths. Insight changes succession to simultaneity, and the pieces of one's personal puzzle fall into place. Bachelard (1939) while speaking of poetry, called certain moments 'rapture' and 'ecstasy' where the person, author, or reader is compelled to leave the continuum of the world's time, eliciting instead a move upwards or downwards in a rupturing moment of verticality. This can also be said of the waking dream. The instant in front of a menacing figure or aura of light becomes a sudden burst of consciousness which calls for an acute act of attention. Such moments (or *augunblick*) of transformation awaken us to a sur-reality beneath and beyond ordinary notions of the real. In the 'a-ha moment' the temporalising horizons of before and after dissolve to reveal a bare inaugural moment. *Kairos* is informed by the emergence of meaning; it liberates us from the ineluctable flow of *durée*.

The most significant events in a person's psychological existence are likely to be precisely those which are 'immediate', breaking through the usual steady progression of time, like a sudden insight, or a view of beauty that one sees in an instant (a marriage proposal etc.) but which may remain in one's memory for days, months, and even years. Jung, in *Memories, Dreams, Reflections*, states:

> To-day I can say that I have never lost touch with my initial experiences. All my works, all my creative activity, has come from those initial fantasies and dreams which began in 1912, almost fifty years ago. Everything that I accomplished in later life was already contained in them, although at first only in the form of emotions and images.
>
> (Jung 1963/1995: 217)

One of the debates in RED as well as in active imagination is whether the experience of the waking dream itself is therapeutic. Jung often emphasized the fact that it is not always necessary to conceptually understand the images coming from the unconscious in order for them to be efficacious for the maturational process, but he always insisted on the primary importance of the *lived experience* of these images: 'Creative processes can be truly grasped only by experience, though intellect may give them a name . . . Experience not books, is what leads to understanding' (Jung 1968: para. 564).

Nonetheless, as to waking dreams, Jung emphasised the need to draw again the imagined experience and to keep on looking at them until meaning is elicited. So, for Jung the spending of time on the imagined experience is crucial for insight and understanding since, as he explains in *Memories, Dreams and Reflections*, if we do not 'understand their meaning they remain a diabolic mixture of the sublime and the ridiculous' (Jung 1963/1995: 202). The post-Jungian James Hillman in *The Dream and the Underworld* (1979) disagrees with the obsession of analysts to bring to light from the underworld the unconscious images and to give them some form of interpretation, since dream images are an initiation of the ego in the psyche.

On the Desoillian side, the oneiric drama itself can also lead to sublimated feelings. For sublimation to be achieved, Desoille urged the patient to explore the heights of the imaginative landscapes since the images encountered there are very powerful and positive. Desoille suggested to his patients to take an upward direction with the help of a staircase or by climbing a mountain or by flying, since these actions transformed the waking dream into an imagination of movement.

Nonetheless, despite emphasising the importance to the affective catharsis and sublimation of the waking dream itself, Desoille also devoted time to go over the imagined scenario and to explore what it could mean for the patient in a separate face-to-face session. The patient would have to write his own personal mythopoeisis of the imaginative journey of initiation from the highest heights and the lowest lows. In the next session, patient and therapist would look again at the waking dream content and compare it with the patient's difficulties in his daily life. During this part of the session, patient and therapist would sit face to face (*vis-a–vis*). Desoille's ideas about interpretation can be paraphrased as always useful, sometimes necessary, but never indispensable.

Kairos: the age of cyberspace offering opportunities to update waking dream methods

One may wonder on the relevance of waking dreams as psychotherapeutic methods which were developed in a modern era to a contemporary audience obsessed with Virtual Reality (VR), a touch screen away. As mentioned above, at the beginning of *Liber Primus* of *The Red Book*, Jung (2009: 229) contrasted the 'spirit of this time' with the 'spirit of the depths' and had to follow the latter so as not to lose himself. I wonder: Can the spirit of the times, namely the age of cyberspace, facilitate our engagement with the depths? It appears that Virtual Reality can offer a way to update waking dreams.

Cyberspace can, in Jungian terms, allow us to explore different aspects of our unconscious and can become a playground and a portal for the unconscious; it can thus offer us a form of active imagination. Browsing the internet freely without any specific goal can also serve as a portal to one's own unconscious and can prove to be a contemporary form of automatic writing. In cyberspace we lose our specific subjectivity, and identities are effortlessly switched off. Nowadays, youths increasingly identify with avatars which they construct as their self-image. As psychoanalyst Sherry Turkle observed: 'The Internet has become a significant social laboratory for experimenting with the constructions and reconstructions of self that characterize post-modern life, we self-fashion and self-create' (Turkle 1995: 180).

Jung was not so keen on modern technology. He believed technology cuts us off from our primitive nature and our instinctual life, and leads to a destruction of the earth. For Jung, the world was no longer enchanted. Yet, Virtual Reality is the ultimate form of high-tech, moving from Weber's (1919) *disenchantment of the world* to Lyotard's (1994) *technological sublime* state. The virtual, in fact, is not the opposite of the real, far from it: It is a mode of being 'fecund and powerful', which allows the blossoming of creation and opens up mines of meaning under the 'platitude' of the immediate physical presence (Lyotard 1994: 10).

Furthermore, with the advent of new techniques for generating, distributing and presenting images, the computer has transformed the image and now suggests that it is possible to enter it (Popper 1995: 3). Banos and Botella (1999) explain how VR can be used for experiencing different identities and even other forms of self. VR applications can be developed for Transactional Analysis, Gestalt and Transpersonal therapy. There is no reason why it cannot be also applied to Jungian active imagination and RED therapy. Virtual Reality could be used for simple hypnotic regressions, re-parenting, the Gestalt two chair technique or dialogue with figures like the shadow, anima/animus or the wise self. Psychodrama scenarios with role playing and other players could also be useful.

Cyberspace seems to provide a good medium to present and market traditional waking dream methods to a more contemporary audience. It is up to practitioners to take up the challenge, this *kairotic* moment, in a world which is increasingly being taken over by ego and supportive therapies. I thus propose a 're-awakening' of waking dreams in therapeutic practice by updating them and making them more relevant to the *spirit of our contemporary times*, through the medium of cyberspace and Virtual Reality.

Conclusion

A waking dream can help us to re-live the past in the present and to anticipate our future. Waking dreams provide us with a different experience of time and thus help us to work through our complexes. Like Jung's Word Association Test, waking dreams help us to access the emotionally toned complexes, as they manifest in different unconscious figures and beings. They also give us an indication of the temporal development of our complexes whether in a pre-verbal or a verbal stage of development. In experiencing unconscious material through the waking dream, insight is possible within the waking dream's time frame as well as after the oneiric experience is over. These insights are experienced more as a revelation, as chronological time is upended and reversed, where past and future are re-inscribed in a timeless moment. The unravelling of one's mythopoesis within a waking dream leads to moments of insight and *kairos* related to one's forgotten past, while offering us leads into the future.

Notes

1 '"En avant" (anticipation), "en haut" (ascenscion) et "a cote" (polarisation). L'evolution existentielle est ainsi un processus complexe qui implique le fait simultane' d'anticiper le future par des anticipations precises, de 'monter; vers les niveaux surordonnes de conscience et de polariser les oppositions caracteristques du monde dans lequel vit l'etre humain' (Berta 2000: 12).

2 In Greek mythology *Kairos*, son of Oceanus and brother of Nike is a winged young man holding a scale on the razor's edge, an allegorical image of the fickleness of his favours, which one must know how to seize 'on the fly'. Opportunity later came to be represented as a maiden with her face covered by a thick shock of hair and the back of her head completely bald. Man must show his skill by grabbing the lock of hair before the girl turns and vanishes forever (Battistini 2005: 16).

3 These anxiety-reducing techniques of Desoille's RED today resemble more Wolpe's use of systematic desensitisation, especially in the forms developed by Lazarus, using 'emotive imagery' in which the subject imagines himself engaging in activities with competence and success, as a means of reducing anxiety.

References

Alexander, F., and French, T. M. (1946). *Psychoanalytic Therapy*, New York, Ronald Press

Bachelard, G. (1939/1966). Instant poetique, instant metaphysique, in *L'intution de l'instant* [The poetic instant and the metaphysical instant, in *The instant intuition*], Paris, Gonthier

Balzarini, G. (1985). L'analisi immaginativa, intervista al Dr G. Balzarini [Imaginative analysis, interview with Dr G. Balzarini], *Synthesis*, IV (7), Piovan Editore

Banos, R. M., and Botella, C. C. P. (1999). Virtual reality and psychopathology. *Cyberpsychology and Behaviour* 2: 283–92

Battistini, M. (2005). *Symbols and Allegories in Art*, trans. S. Sartarelli, Los Angeles, California, Getty Publications

Bergson, H. (1960 [1888]). *Time and Free Will: An Essay on the Immediate Data of Consciousness*, trans. F. L. Pogson, New York, Harper and Row

Berta, M. (2000). *Les images qui guerrisent. Le Rêve Eveillé Dirigé de Robert Desoille*, Montevideo, unpublished manuscript

Cassar, L. (2014). Shades of RED: Jung's Technique in the Red Book and Desoille's R.E.D (Rêve éveillé dirigé) Method. In E. Kiehl (ed.), *Proceedings of the X1Xth International Congress for Analytical Psychology – 100 years on: Origins, Innovations and Controversies*, Copenhagen 2013, Einisedeln, Daimon

Desoille, R. (1938). *Exploration de l'affectivite subconsciente par la méthode du rêve éveillé: sublimation et acquisitions psychologiques* [An exploration of the subconscious emotions through the waking dream method: sublimation and psychological acquisitions] Paris, J.L.L. D'Artrey

——. (1945). *Le Rêve Éveillé en Psychothérapie: Essai sur la Function de Regulation de l'Inconscient Collectif* [The waking dream in psychotherapy: essay on the regulatory function of the collective unconscious], Paris, Presses Universitaires de France

——. (1955). *Introduction a Une Psychothérapie Rationelle* [An introduction to a rational psychotherapy], Paris, Editons l'Arche

——. (1961). *Theorie et Pratique du Rêve Éveillé Dirigé* [Theory and practice of the directed waking dream], Genève, Éditions du Mont Blanc

Freud, S. (1908/1972). The relation of the poet to day-dreaming. In P. Rieff (ed.), *Character and Culture*, New York, Collier Books

Hillman, J. (1979). *The Dream and the Underworld*, New York, Harper and Row

Jung, C. G. (1911–12/1952). *Symbols of Transformation*, CW Vol. 5

——. (1921/1971). CW Vol. 6, *Psychological Types*

——. (1935/1976). *The Tavistock Lectures*, CW Vol. 18

——. (1951/1968a). *The Psychological Aspects of the Kore*, in CW Vol. 9i

——. (1951/1968b). *The Psychology of the Child Archetype*, in CW Vol. 9i

——. (1953/1968). Epilogue, in CW Vol. 13

——. (1955/1970). *The Transcendent Function*, in CW Vol. 8

——. (1963/1995). *Memories, Dreams, Reflections*, London, Fontana Press

——. (1973). *Letters (1951–1961), Vol. II*, Princeton, New Jersey, Princeton University Press

——. (1998). *Visions: Notes of the Seminar Given in 1930–1934 by C. G. Jung*, C. Douglas (ed.), London, Routledge

——. (2008). *Children's Dreams: Notes from the Seminar Given in 1936–1940*. L. Jung and M. Meyer-Grass (eds), Philemon Series, Princeton, New Jersey, Princeton University Press

——. (2009). *The Red Book, Liber Novus*, S. Shamdasani (ed.), trans. M. Kyburz, J. Peck, and S. Shamdasani, New York, W.W. Norton and Company

Kluft, R. P. (1989). Playing for time: temporizing techniques in the treatment of multiple personality disorder. *American Journal of Clinical Hypnosis* 32(2): 90–98

Lewis, C. S. (1950). *The Lion, the Witch and the Wardrobe*, London, Geoffrey Bles

Lyotard, J. F. (1994). *Lessons on the Analytic of the Sublime*, trans. E. Rottenberg, Stanford, California, Stanford University Press

Popper, F. (1995/2007). *From Technological to Virtual Art*, Cambridge, Massachusetts, MIT Press

Reddemann, L. (2008). *Psychodynamisch imaginative Traumatherapie. PITT – Das Manual* [Psychodynamic imagery trauma therapy. PITT – The manual], 5th ed. Stuttgart, Klett, Cotta

Sells, M. (1996). *Early Islamic Mysticism*, New York, Paulist Press

Smucker, M. R. (1997). Post-traumatic stress disorder. In R. L. Leahy (ed.), *Practicing Cognitive Therapy: A Guide to Interventions*, 193–220, Northvale, New Jersey, Jason Aronson

Turkle, S. (1995). *Life on the Screen. Identity in the Age of the Internet*, New York, Touchstone

Virel, A., and Fretigny, R. (1968). *L'imagerie Mentale: Introduction a l'onirothérapie* [Mental imagery: An introduction to oneirotherapy] Genève, Éditions du Mont Blanc

von Franz, M. L. (1980). On active imagination. In I. F. Baker (ed.), *Methods of treatment in Analytical Psychology – VII International Congress of the International Association for Analytical Psychology*, 88–99, Fellbach, Bonz

Weber, M. (1919/1946). Science as a vocation. In H.H. Gerth and C. Wright Mills (eds and trans.), *From Max Weber: Essays in Sociology*, New York, Oxford University Press

Yiassemides, A. (2014/2016). *Time and Timelessness: Temporality in the Theory of Carl Jung*, London and New York, Routledge, Taylor and Francis Group

5

THE THIN CURTAIN OF NON-SPACE AND NON-TIME

Synchronicity and prospective dreams; theory and clinical applications

Angeliki Yiassemides

In the opening scene of Asimov's 'The End of Eternity' (1955/2000), Harlan, Asimov's protagonist, stands in front of the thin curtain of Non-Space and Non-Time which separates eternity from ordinary time. Harlan is an Eternal: he is not really concerned with the concept of time in the way that we mortals – or 'Timers', as Asimov called us – are. Yet, even though he is accustomed to the parallel existence of different temporalities along-side the Eternal realm, he, too, must consult the Temporal Handbook. This is how complicated temporality is: Causality, temporal linearity, eternity, timelessness, relativity, Chronos, Kairos, all belong to the 'temporal handbook' that we humans have been writing about, since the beginning of Time.

Some theories are bound by one kind of temporality; most commonly causality. This does not hold true for Analytical psychology, which assumes that a principle other than causality must be in operation in nature. According to Jung's theory, the principle of synchronicity is the 'added principle' to causality that must be considered if we are to adequately (attempt to) understand the psyche, and the cosmos at large. Synchronicity suggests that there are moments in Time when we can look through 'the thin curtain of Non-Space and Non-Time' and get a glimpse of a different temporal arrangement of events, beyond ordinary Time.

In this chapter I will be presenting some clinical examples, alongside theoretical considerations regarding the principle of synchronicity that point towards a different temporal understanding. It is my hope that the reader will find such elaboration useful in his daily life, and that the clinician will benefit from the exposition of the material which is steeped in unorthodox temporal order.

A falling together in time

Jung's devotion to synchronicity led to the development of his theory and to a deeper understanding about the working of, and the interconnection between, the psyche and the world. This deems synchronicity one of the most important phenomena that demonstrates the nature of the psyche and the cosmos as described by Jung:

> [Synchronicity] cannot be a question of cause and effect, but of *a falling together in time*, a kind of simultaneity. Because of this quality of simultaneity, I have picked on the term 'synchronicity' to designate a hypothetical factor equal in rank to causality as a principle of explanation.
>
> (Jung 1952: para 840)

What were the chances that a scarab would tap on Jung's window at that exact moment in time when his client was narrating her dream of a golden scarab?[1]

As Jung pointed out, this – as all synchronistic incidents – cannot be understood by causality: The dream did *not* cause the beetle to tap on the window; and certainly, the insect did not cause the dream. Nonetheless, these two events – the inner event of a dream and the outer event of the insect at the window – were meaningfully connected for Jung and his patient. And for meaning to be attributed by both observers these two events had to happen *around the same time*. As Jung put it, this was a 'falling together in time.' Had these events happened separately in time, say for example the beetle tapping on the office window two weeks later, these events would not have been meaningfully connected by Jung or his patient.

As I have emphasized elsewhere (Yiassemides 2011; 2014/2016), the workings of time seem to be at the core of the principle of Synchronicity; and that is why Chronos (the Greek word for Time) is embedded in the term. The time factor in synchronicity has been criticized or at best downplayed, but I believe it is vital if we are to understand the essence of the principle as intended by Jung in his formulations.

Time's flow and the attention given to the events that happen *within each time-frame* is essential. The manner, in other words, events 'fall together in time' is critical in attributing meaning to them and classifying them as synchronistic. This understanding of time implies a kind of *psychic 'unit' of time* that is different – in its parameters and definition – from a unit of physical/real time (i.e. clock time). This is intimately associated with the concept of relative time – as opposed to a fixed and universal 'real time' – in the sense that each observer of Reality creates her own experience depending on 'where' she is.

For Jung, the psyche has a *temporally relative nature at large*; timelessness is not contained in the unconscious. The emergence of consciousness does not create an opposing 'kind' of time (as in Freud's theory). For Jung, the psyche and the cosmos at large, in all their manifestations, are at their core *timeless*. Time in the psyche, as well as in the outer world, is relative. This implies an understanding that the psyche contains 'the future' in 'the present' moment (Yiassemides 2014/2016).

Synchronicity can help us 'see' the workings of time and timelessness, in 'real-time', as it were. This concept is intimately related to our temporal experience; and that is true of Jung's own experience and the way his overall theory unfolded. Particularly, Jung's conviction in a *non-traditional temporal frame*, his understanding regarding the interconnection of time and timelessness, and his acceptance of the oneness in nature allowed for synchronistic experiences, and other similar inexplicable-by-causality events, such as telepathy and foreknowledge, to be experienced and to be considered plausible and even probable.

Jung described synchronicity as 'the simultaneous occurrence of a certain psychic state with one or more external events which appear as meaningful parallels to the momentary subjective state – and in certain cases, vice versa' (Jung 1952: para 850). Therefore, even in cases where the observable events that are meaningfully related might not happen simultaneously in clock time, they are experienced as happening in a particular *momentary subjective state*. I have proposed that, 'this can be read to mean a certain phase in one's life, hence the characterization of it being subjective (i.e. belonging to/taking place in one's mind, rather than being objective)' (Yiassemides 2014/2016: 53). These subjective temporal parameters correspond to the 'psychic units of time' as I have suggested above, but do not necessarily correspond to external time. 'At such times, something – i.e. a theme, a pattern, an issue, a constellated archetype – is active so that one notices certain types of experiences which correspond with (or are parallel to) certain outer events' (Ibid.).

This explanation is in accordance to Jung's formulations, first presented in 1929: 'I have invented the word synchronicity as a term to cover these phenomena, that is, things happening at the same moment as an expression of *the same time content*' (Jung 1984: para 417, emphasis added). Outer and inner events are connected in time since they belong to 'the same time content' of the observer's life, i.e. of the time-frame of the one who attributes meaning to the acausally connected events.

Prospective dreams

Accepting the assumption that there is a non-causal principle in operation can lead to a different experience of time and life itself; and this can be extended to the analytic situation where such an understanding can enrich our experience, both as analysts and analysands. In what follows, I will focus primarily on a particular kind of synchronicity: future-telling dreams, or more accurately *prospective dreams*. Jung welcomed prospective dreams and visions and found them to be of great relevance. Prospective dreams are one of the possible manifestations

– or perhaps 'types' – of synchronicity where the workings of the principle are clearly illustrated. Causality is negated by default since the event dreamed by the dreamer has not *yet* occurred in 'awaken life'. Thus, this kind of dream exemplifies Jung's understanding and illustrates the workings and manifestations of temporality in the psyche.

Jung's belief that such dreams can take place allowed for the development of his theory regarding the psyche's temporal manifestations and provided suitable examples for the development of his theory. His overall theoretical frame allowed him to expand the traditional world-view and to reach beyond determinism. He recorded several instances of prospective dreams in his career, both from his personal life as well as for his clients' lives.

Prospective dreams 'foretell' what is to happen in the future. They are a clear example of what I would call an *unorthodox temporality in action*. Such dreams are 'an anticipation in the unconscious of future conscious achievements, something like a preliminary exercise or sketch, or a plan roughed out in advance' (Jung 1916/1948: para 493). Jung did not find the term 'prophetic' suitable for such dreams: 'It would be wrong to call [prospective dreams] prophetic, because at bottom they are no more prophetic than a medical diagnosis or a weather forecast' (Ibid.). Such dreams, in which the dreamer is given *an indication of the future* do not agree necessarily in every detail with waking life, and so cannot be called 'prophetic'.

The future: beyond time-traveling

Jung was able to gaze and dwell in this alternative temporal Reality that existed beyond ordinary understanding: As far as Analytical Psychology is concerned, 'the future' does *not* signify a specific point in time down the linear temporal line; 'the future' is not a particular outcome that waits for us in time. Such 'future' would imply a concrete section/point in time where we will eventually arrive at, as we are traveling through linear time – like the man traveling back to kill his own grandfather, only to find out that he no longer exists in the future. This 'static' view of the future is what we usually encounter when we talk about time-traveling as a form of movement along the linear temporal highway.

Jung's position regarding the future dimension is in no way related to linearity, nor is it associated with causality. And this is in line with Jung's overall temporal framework which goes beyond causality and favours a non-linear, non-causal approach. The principle of teleology (or 'finality' as Jung came to call it) can accommodate for occurrences such as precognition. Jung believed that causality and teleology are two complementary, i.e. mutually exclusive, and equally necessary methodologies for the complete interpretation of reality. Teleology goes beyond the end results by considering the 'living meaning' of a process (like the Aristotelian Final Cause which explains processes by means of their final expression).

I have suggested elsewhere that the future directionality of the psyche is what Jung attempted to capture by the adoption of teleology as an additional principle to causality: 'Teleology implies future-oriented movement to a point in time where (ideally) a more improved/ integrated "version" of the psyche lies. Even in the event of regression (i.e. a blatant movement backwards) there exists a teleological, future-directed point of development' (Yiassemides 2014/2016: 29). To begin to understand Jung's teleology we must break away from concrete temporal demarcations and causality as the only possible explanation to reality, and employ a viewpoint where the future, the past and the present have *symbolic dimensions* that provide the psyche with a dynamic movement forward. 'The psyche, due to its "teleological impulse" is inclined towards futurity, but not towards a concrete future. Telos is *not* based on linear temporality, but rather it is activated now. Not "now" as in the "present moment," but now as the force that is constantly present and gives vitality and movement to the psyche' (Ibid.).

The future, for Jung, is a *dynamic dimension* that constantly influences the past and the present. There is no pre-fixed telos, i.e. an end point, which could be foretold; the past does *not cause* the future. Jung spoke of 'an anticipatory combination of probabilities'; these probabilities are not concrete upcoming outcomes and that is why they do not coincide in detail with what will unfold in awaken life; rather, these *probabilities* 'may coincide with the actual behaviour of things' (Jung 1916/1948: para 493). It is as if the unconscious provides the dreamer with an array of probable outcomes based on data provided by 'perceptions, thoughts and feelings' as well as 'subliminal memories' (Ibid.).

Consciousness, Jung believed, is lagging behind the unconscious, and that is why prospective dreams are possible. The psyche offers information that has *not yet registered in consciousness*, and thus '[t]he prospective function of dreams is sometimes greatly superior to the combinations we can consciously foresee' (Jung 1916/1948: para 493). Jung spoke of the 'feeble accentuation' between various contents of the psyche (such as the above-mentioned aspects of perceptions, thoughts, feelings and subliminal memories): Even though these elements are not connected in a strong enough way – as far as consciousness in concerned – they are nonetheless connected and influence our psychic life. It is, therefore, reasonable that 'with regards to prognosis . . . dreams are often in a much more favourable position than consciousness' (Ibid.).

Moreover, similarly to all kinds of dreams, not all prospective dreams belong to one's personal unconscious; a prospective dream could be drawn from the collective unconscious, to which each individual has access. For Jung, there are precognitions that belong to a collective event; he called these 'Big Dreams'. He provided several examples of dreams that seem to predict the future, that lie outside the personal world of the dreamer, as for example his 1913 vision and recurrent prospective dreams of the First World War.[2] Jung's psyche, being part of the collective whole, had access to a reality beyond his personal circumstances and knowledge; he had access to his epoch's past, present and future, all of which coexisted in the realm of the collective, *at that particular 'momentary subjective state.'*

For Jung, instances such as premonitions, second sights, clairvoyance and foretelling dreams are examples of the 'long-range effects in time' (Jung 1897: para 131) – in other words, of the psyche's ability to extend far into the future. In his *1936-1940 Seminars on Children's Dreams*, he used the example of prospective dreams to exemplify the psyche's *transcendence of time*. He observed that dreams dreamed by children can reveal details regarding 'crucial future events' that 'are anticipated in a surprising way' (Jung 2008: 18). The totality of our personality, Jung explained, is present at the onset of our lives and that is the reason why such dreams take place in childhood. It appears illogical that such dreams can anticipate the child's developmental process into adulthood since such futuristic images of one's complete, as it were, state cannot pre-exist. Nonetheless, 'The individual is what it will be . . . Later, life forces us to make one-sided differentiations. But that is why we get lost to ourselves and must learn, again, to find ourselves. When you are whole, you have discovered yourself once again, and you know what you have been all the time' (Jung 2008: 20). Such prospective dreams are the product of the psyche's non-linear, non-causal temporality. The 'future' already exists in the 'present'.

Jung was not suggesting that our lives are pre-destined. He was referring to a *future dimension* that rests on an understanding of *relative time*; the future has a relative meaning in regards to a linear view of time. Future potentialities cannot be understood under a causal paradigm. Causality, Jung wrote, overlooks the 'final significance' of events. He emphasized the importance of the events' function and *future direction*: 'It is impossible to consider the psyche from the causal standpoint only; we are obliged to consider it also from a final point of view' (Jung 1916/1917: para 688). We must employ a non-causal arrangement to conceptualize this phenomenon and accept the possibility that in the future we could rediscover our wholeness which was present since 'the past'.

If we employ causality, we are constrained by a causal temporal understanding of time-traveling, whereupon we would need to suggest that the dreamer is somehow capable of visiting a particular future point while asleep, akin to the grandchild killing his own grandfather. However, according to a non-causal/acausal temporal paradigm, the past, the present, and the future are *not* sequentially arranged and, thus, the future is not perceived as a literal end-of-the-line state. The future (just like the past and even the present) is an *abstract dimension* that exists in the present and existed in the past. Indeed, all temporal demarcations (that characterize linear time upon which causality operates) coexist at any given moment: *everything is simultaneous and contemporaneous.*

Moving away from causality: the future dimension

In Asimov's science fiction novel, *The End of Eternity*, the world is arranged in a hierarchy of temporal beings. Above the Timers who are trapped in a 'primitive' sense of time, there are the Eternals who are responsible for temporal tampering in the mortal dimension. And above them there are the agents of Reality Change. Each hierarchically arranged kind of being is able to view a different kind of temporal reality. Timers, like ourselves, can only view one temporal reality at a time. And whereas the Eternals believe there are many Realities but only one exists at a time, in the eyes of superior temporal beings several Realities coexist simultaneously and are seen in their state of non-Reality. Asimov's world was ahead of its time: future realities are *possibilities unfolding onwards*, at a future dimension. And so were Jung's postulations regarding prospective dreams, synchronicities, and a non-causal arrangement of the cosmos at large. And yet, his methodological understanding was not presented in a vacuum. Jung (and most certainly Asimov as well) was largely influenced by 'the spirit of the times', that is, by the overall scientific developments of his era. He observed scientific developments with fascination and was in contact with important figures in the field of physics, such as Albert Einstein, Wolfgang Pauli, and Werner Heisenberg among others, who questioned the role of causality and determinism.

Following the lead of these scientists, Jung steered away from the traditional approach widely employed by the great majority of his contemporary theorists (including Freud). He rejected the exclusive use of causality and suggested instead a different way of explaining the arrangement of events which would allow for the exploration of *the psyche's totality*, and its place in the cosmos at large.

His teleological understanding allowed for non-causal events to be considered as equally acceptable interpretations of reality, in a similar way that Einstein's theory of relativity – which was presented prior but during the same era as Jung's theory of synchronicity – complemented the existing deterministic parameters of the universe. Albert Einstein, one of Jung's contemporaries, was the physicist who, as Jung put it 'first started me off thinking about a possible relativity of time as well as space and their psychic conditionality' (Adler 1974: 109). Jung acknowledged that his interactions with Einstein, 'exerted a lasting influence on my own intellectual work' (Ibid.). Even though Wolfgang Pauli was the most influential scientist for Jung's theory,[3] it appears that Einstein's General Relativity is the theory that best supports this particular temporal aspect of Jung's understanding of prospective dreams. General Relativity is in agreement with the *simultaneous and contemporaneous coexistence of temporal demarcations*, as supported by Jung's theory. In Einstein's universe, *the future is in existence.*[4]

Einstein across the river: Marta's dream

On our initial phone call, five years ago, Marta wanted to enter analysis because she felt trapped, she could not move forward; she needed to leave her husband's 'home' (she was at the time an employee at his business), and to leave behind her abusive paternal family. As a child, Marta lived through the Turkish invasion of Cyprus in 1974 during which her family lost everything. She, like many refugees after the war, experienced extreme poverty; her marriage was a way out of the struggle for survival. Marta could not find joy in life. She was having panic attacks and felt very vulnerable; she had suicidal thoughts, and could not see the point of living in fear anymore. The future was unbearable and unimaginable. She told me that she had to find the courage to leave her job or she would die, but she was afraid of her husband's reaction, which was equally murderous.

It was Marta's first session. We had spoken on the phone the previous week and I had given her directions to my office. She arrived there after nightfall: It was already dark, and she was thus unable to understand where exactly the office was in relation to the greater area. She told me a dream she had the night before she had come to see me, which she was reading from her dream journal:

She is at the end of a dead-end street, at the end of which there is the river bank. She described in detail that street: at its end, there is a landmark building (i.e. particular building, with a particular name). Outside that building there is a fair with food stands, a festive event. She is wandering about; she sits on the pavement towards the end of the road near the river bank, not knowing what to do next. There is a man who is trying to communicate something to her. She describes him as someone who looks like Albert Einstein. He isn't talking. Instead he is pointing the way: He is showing her the other side of the river bank. She cannot understand what he means, and assumes at the time, as she is re-telling the dream which she is reading it out loud from her diary, that he is referring to the river itself; that she must go to the river bank.

Marta left that evening unaware of the fact that she had had a prospective dream, a dream that foretold the next day's events. The next time she came to my office it was daytime. I opened the door to find a shocked woman. Daylight revealed the missing details that were needed to make sense of her reported dream: My office is located at the end of a dead-end street, near a river bank, *directly across* from the street where her dream's events unfolded: in the daylight she could see from outside my office, across the river bank, the landmark building she had dreamed of the previous week. She realized that in her dream, Albert Einstein showed her the way to my street *the night before* our first session.

I approached this dream using Jung's time theory that points the way to a multi-directional multi-dimensional temporal reality. While her consciousness rested, Marta gained access to *the totality of the psyche*: a totality that included her past (her inability to move forward, her reaching of the dead-end), her present state (her decision to enter analysis), and her future (our work together 'across the river bank') all within one dream. Her psyche, gravitating towards the future, pointed the way forward, before the events unfolded in linear time.

I was aware of the prospective nature of her dream an entire week before Marta (if you recall, she did not realize where my office was located until a week had passed, when she came for our second session during daylight). There is a further, multi-dimensional aspect to this dream, since beyond the obvious prospective aspect for Marta it was, at the same time, a synchronistic occurrence for me, her therapist:

At that time, I was working on my PhD where I was focusing on Einstein's theory of relativity. I was trying to comprehend and find examples of this elusive yet profound theory. The implications of Einstein's theory on temporal dimensions and the abolition of linearity were of particular importance to my work. I was wondering and pondering what the parallels and metaphorical consequences were for the psyche and the cosmos at large. I found Marta's dream astonishing. What she was narrating was inconceivable from a linear point of view. If I was to understand what she had experienced – a dream predicting the near future – I had to embrace the relativity of time as a *reality of the psyche*, surpassing theory.

The 'observer' of the synchronistic occurrence was not simply the patient, but also myself the analyst who was at the time trying to grasp the temporal elements of relativity, and trying to find ways to incorporate them in Jung's principle of synchronicity. Marta's dream helped me create the necessary frame within which to better digest and clinically utilize the paradoxical temporality I was studying at an intellectual level. Her dream served as the 'grounding event' which brought together the intellect and the clinical pieces of my work.

This was such a clear illustration of what I was struggling to understand at the time, that it has become a vital part of my experience of synchronicity in my practice; it provided a context within which to understand *the acausal nature of the psyche*. Beyond the dreamer – that is, the 'producer' and 'recorder' of the prospective dream – I, the therapist, the 'witness of the dream', was also a meaning-making observer. Meaning was not unidimensional: it was layered and had multiple 'observers' and 'meaning-makers'.

The 'observers' in this case were two (like in the Golden Scarab example given by Jung), so the event was deemed synchronistic by *both* myself (the analyst) and the patient. However, for each one it was a different 'synchronistic reality'. As an analyst, I was the witness of *her* synchronicity (i.e. her dream that predicted the location of my office which she was to visit the following day). But, equally significant, I was also the observer of another level of reality: the synchronicity that occurred in my *personal* field, in *my individual momentary subjective state*. My client's dream was acausally related to what I was preoccupied with at that subjective time of my life (i.e. Einstein's theory of relativity and its transcendence of linear time). For Marta, this was a strange prospective dream that confirmed her decision to begin analysis, and paved her way into this elusive yet observable field. She had a dream that had prepared her for the linear unfolding of events – as Jung put it, in such dreams 'the unconscious anticipates things' (Jung 2008: 11).

This temporal directionality, which is intimately related to Jung's teleological understanding, is an important and unique element of his psychoanalytic approach. In Jung's theory, the future dimension holds a privileged position: the psyche which is *relative at large*, is characterized by a *purposeful and forceful movement forwards*. We can find elements of this temporal directionality in his description of the archetypes which are timeless and belong to the eternal, but at the same time have a certain temporal direction: they develop in time and *through generations*, they 'have crystallized out in the course of time' (Jung 1917/1926/1943: para 151). Archetypes, which originate in our ancestral past, have a future orientation. Here lies an important distinction between Freud's and Jung's temporal understanding; in Freud's theory, there is disproportionate attention paid to the past and a noticeable absence of the future dimension (which explains Freud's rejection of 'prophetic dreams': for Freud, the future-telling aspect of the dream is a mere projection of the wish).[5]

For Jung, the psyche 'grounds' itself to the past; at each new stage of our development it is essential to 'orient to the past' in order to reconnect with our inherited origins:

> If we cannot deny the archetypes or otherwise neutralize them, we are confronted, at every new stage in the differentiation of consciousness to which civilization attains, with the task of finding a new interpretation appropriate to this stage, in order *to connect the life of the past that still exists in us* with the life of the present, which threatens to slip away from it. If this link-up does not take place, a kind of rootless consciousness comes into being no longer oriented to the past, a consciousness which succumbs helplessly to all manner of suggestions, and in practice, is susceptible to psychic epidemics.
>
> (Jung 1940: para 267, emphasis added)

We need to acknowledge the past, and to continuously update our connection with it, by reinterpreting it. And just like 'the future' which is a *dimension* rather than a concrete point down the linear time-line, 'the past' is seen as a past dimension of the psyche, as opposed to a concrete, historical point in the psyche's past. The past is constantly active and must be *modernized* in the present moment, with which it constantly interacts: 'The most we can do is to dream the myth onwards and give it a modern dress' (Jung 1940: para 271). This implies a dynamic and active temporality within which the psyche operates *at all times*. The function of these dimensions, i.e. their teleological significance, is to give meaning in the present by connecting the 'now' with the past.

Seen from a *temporally relative standpoint*, all temporal demarcations – the past, present and future – are in constant interaction 'now', literally 'at any given moment'. In other words, Jung's temporal multidimensionality is characterized by temporal *multidirectionality*, which flows dynamically and purposefully towards the future dimension: 'Life is a flux, a flowing into the future, and not a stoppage of backwash' (Jung 1940: para 278). The psyche's 'teleological aim' is towards the future, towards its self-realization. For the process of individuation to occur – which has as its goal 'the synthesis of the self' (Ibid.) – the entire temporal spectrum must be considered: wholeness is achieved by the *unification of the opposites* that are embedded in the past and the future dimensions.

> On the one hand [the psyche] gives the picture of the remnants and traces of all that has been, and on the other, but expressed in the same picture, the outline of what is to come, in so far as the psyche creates its own future. The psyche at any given moment is on the one hand, the result and culmination of all that has been and on the other, a symbolic expression of all that is to be.
>
> (Jung 1914: paras 404–5)

Lost and found: Marta's journey

Marta's work since the very beginning has been accompanied by regular synchronistic occurrences both in prospective dreams and events in her awaken life. At times, just like in the dream above, the synchronistic occurrences that unfold have multiple levels. They are meaningful, not just for Marta but also for me, her therapist. These events point the way forward for the therapeutic process by reaffirming to both the analysand and the analyst that the future dimension is unfolding in a mysterious yet observable way. As I was putting together this chapter one such event took place within the analytic frame. This synchronistic event helped me to 'frame' this chapter within the theoretical parameters presented above, and provided a bridge to the concrete and clinical.

Marta had a particularly upsetting dream and was devastated by a feeling of rage regarding events that unfolded in her life; she was overwhelmed with negative emotions. She sat on the armchair unable to contain herself. She used a Greek expression – 'space is not big enough to contain me' (*'δεν με χωρά ο χώρος!'*) – to describe how uncontained she felt. She repeatedly declared herself 'lost'. At that moment in time I thought to myself that I needed to find a way, within the time-frame of the session, to help Marta get back on track, indeed to 'find herself', and to find a way to connect within. We sat in silence.

In that silent space, I was contemplating bringing back an image presented in a dream she had the previous week, during which she was traveling alone, barefoot on her own path, towards her home by the sea. I was thinking that this image – which spontaneously arose in my mind – was perhaps a helpful one for Marta at this particular moment in time.

As we sat in silence, Marta was trying to make herself comfortable on the armchair. Others sat on it since I last saw her, so she was readjusting the pillow to her liking. Then I saw Marta putting her hand in the space between the armchair's bottom pillow and the arm-rest and pulling something out. She looked at it. Her face had an expression of shock and disbelief. I had no idea what that item was, since I had never seen it before and never noticed there was something 'hidden' in the armchair. It was a bracelet. It was *her* bracelet: 'It's mine! I didn't notice I lost it!' Then she looked closely, and she said: 'It says something here.' There was a tiny gold plate hanging from the bracelet. She looked for her glasses, put them on, and with tears in her eyes she read – as if for the first time – the inscription on the lost-and-found bracelet: '*My Journey*.' We stayed in silence once more.

We then spoke about how her anger had taken her away from her inner work, indeed from 'her journey' and how she had a deep desire to return to her path, to regain access to herself. But she could not find a way there. Not until she was reminded that she has an inner journey to return to; there was an inner space that was safe and could help her regain a sense of self.

In a similar way, Marta's dream as presented earlier, had multiple levels of meaning that corresponded to two observers – the analysand and the analyst – this synchronistic occurrence was not only meaningful for Marta, but for me as well: At that moment, I too recalled the 'demand' that Marta made during our first session, several years ago: 'To live without fear, and to go on living *my* life!' The past – her demand of a life free of fear, amidst a fear-full existence and abuse – was brought to the present moment, in the analytic session.

Time collapsed, the spacious present was large enough to include her entire journey: the beginning – which was initiated with the Einstein dream – the path that has unfolded during these years, and this moment in time when we were both reminded, via the synchronistic occurrence, of *her* journey.

Furthermore, this synchronistic occurrence punctuated the feeling of 'lost and found', a sense of regression in Marta's journey. She was finding her path, once more at this 'moment in time', a moment that was much more spacious than 'now' and could encompass the specific subjective temporal frame within which this unfolding was taking place. I felt that she needed to 'regress', as it were, to her inward space, where her trauma lay. And this was a feeling I had had before the bracelet was recovered by Marta underneath the pillow of 'the analytic chair'.

For Jung, the psyche exists above and beyond the physical reality; it is a phenomenon which transcends the realms of time and space; the actual underlying world process, Reality, is beyond linearity. This understanding allowed for the unfolding of a theory that can explain non-linear and non-causal events, such as synchronistic occurrences and prospective dreams. Understanding temporal demarcations as *possibilities* and *dimensions* within a temporally relative order allows for the observation and acceptance of a reality beyond causality and temporal linearity; for a multidimensional and multidirectional arrangement.

Moments such as the ones I have described above, during which the analytic couple work jointly to realize the analysand's authentic future, are often characterized by synchronistic occurrences that are meaningful to both observers. Synchronicity reminds us (both analysts and analysands) that the reality of the psyche is not contained in space, nor time; that the curtain that separates our temporally fixed and linear reality from Non-Space and Non-Time is indeed thin, and can be lifted – even for a moment, long enough to envision a possible future dimension.

Notes

1 'I was sitting opposite her one day, with my back to the window, listening to her flow of rhetoric. She had an impressive dream the night before, in which someone had given her a golden scarab – a costly piece of jewelry. While she was still telling me this dream, I heard something behind me gently tapping on the window. I turned round and saw that it was a fairly large flying insect that was knocking against the window-pane from outside in the obvious effort to get into the dark room. This seemed to me very strange. I opened the window immediately and caught the insect in the air as it flew in. It was a scarabaeid beetle, or common rose-chafer, whose gold-green colour most nearly resembles that of a golden scarab. I handed the beetle to my patient with the words, "*Here is your scarab*"' (Jung 1951: para. 982).

2 'In October, while I was alone on a journey, I was suddenly seized by an overpowering vision: I saw a monstrous flood covering all the northern and low-lying lands between the North Sea and the Alps. When it came up to Switzerland I saw that the mountains grew higher and higher to protect our country. I realized that a frightful catastrophe was in progress. I saw the mighty yellow waves, the floating rubble of civilization, and the drowned bodies of uncounted thousands. Then the whole sea turned to blood . . .' This vision was repeated in two weeks. An inner voice spoke "Look at it well; it is wholly real and it will be so. You cannot doubt it." That winter someone asked me what I thought were the political prospects of the world in the near future. I replied that I had no thoughts on the matter, but that I saw rivers of blood. I asked myself whether these visions pointed to a revolution, but could not really imagine anything of the sort. And so I drew the conclusion that they had to do with me myself, and decided that I was menaced by a psychosis. The idea of war did not occur to me at all' (Jung 1963: 175–76).

3 See Lindorff, 1995; Meier, 2001; Gieser, 2005; Carter and West, 2008; Miller, 2009; Yiassemides 2014/2016.

4 It is perhaps important to clarify at this point that GR as expressed by Einstein's formulations holds a different position on time from QM, which was Pauli's main field of interest. Actually, general relativity and quantum mechanics, the two pillars of modern physics, operate under different assumptions. To this day, 'the most profound question in physics is how to unify the really small [as explored by QM] with the really big', the realm investigated by Einstein's approach (Gefter 2007: 31). Particularly, according to GR time and space are interwoven to form a four-dimensional space–time continuum. Furthermore, *time is not absolute* and even though it is a 'useful measure of things' it is nonetheless 'nothing special' (Gefter 2008: 26). For quantum mechanics, on the other hand, time has an indispensable role since one needs to 'keep track of the ever-changing probabilities that define the microworld' (Ibid.). Thus, an *absolute time* is measured in quantum physics. However, things are not so clear-cut: instead of treating space and time as a whole – which is the case with Einstein's theory of relativity – in quantum mechanics the universe is divided into the quantum system which is under observation and the 'classical world' outside the quantum level of observation; the former measures absolute time whereas the latter views time as relative. It has also been pointed out that the view that 'relativity demands continuity, strict causality and locality [whereas] quantum theory requires non-continuity, non-causality and non-locality' (Arden 1985: 473) is inaccurate: QM is *not* deterministic, at a fundamental level since we never know exactly when decay will occur; we only know the *probabilities* of the various decays.

5 For an elaboration of the differences in Freud's and Jung's temporal directionality in their respective theories, see Yiassemides 2014/2016.

References

Adler, G. (ed.). (1974). *C. G. Jung Letters, Volume II: 1951-1961*, Bollingen Series XCV 2, Princeton, New Jersey, Princeton University Press

Arden, M. (1985). Psychoanalysis and survival. *International Journal of Psychoanalysis* 66: 471–80

Asimov, I. (2000 [1955]). *The End of Eternity*, London, Voyager

Carter, L. and West, M. (2008). Book review of 'the innermost kernel: depth psychology and quantum mechanics; Wolfgang Pauli's dialogue with C. G. Jung'. *Journal of Analytical Psychology* 53: 127–40

Gefter, A. (2007, 9 March). Is the universe a fractal? *New Scientist* 2594: 30–33

——. (2008, 19 January). Time's up. *New Scientist* 2639: 26–29

Gieser, S. (2005). *The Innermost Kernel; Depth Psychology and Quantum Physics. Wolfgang Pauli's Dialogue with C. G. Jung*, New York, Springer

Jung, C. G. (1897). The Zofingia Lectures 83–142: Some Thoughts on Psychology, ed. Sir H. Read, M. Fordam and G. Adler, trans. J. V. Heurck, London, Routledge & Kegan Paul

——. (1914). *The Psychogenesis of Mental Diseases: On Psychological Understanding*, in C.W., Vol. 3

——. (1916/1917) Preface to *Collected Papers on Analytical Psychology*, in C.W., Vol. 4

——. (1916/1948). *The Structure and Dynamics of the Psyche: General Aspects of Dream Psychology* (C.W. 8, pars 443–569)

——. (1917/1926/1943) *Two Essays on Analytical Psychology: On the Psychology of the Unconscious*, in C.W., vol. 7

——. (1940) *Archetypes and the Collective Unconscious: Psychology of the Child Archetype*, in C.W., Vol. 9i

——. (1951) Appendix: On Synchronicity in *The Structure and Dynamics of the Psyche*, in C.W., vol. 8

——. (1952). Synchronicity: an Acausal Connecting Principle in *The Structure and Dynamics of the Psyche*. in C.W. Vol. 8, paras 16–968

——. (1963) *Memories, Dreams, Reflections*. Toronto, Canada, Vintage Books

——. (1984). *Dream Analysis: Notes of the Seminars Given in 1928-1930* by C. G. Jung. W. McGuire (Ed.). Bollingen Series XCIX. Princeton, New Jersey, Princeton University Press.

——. (2008) *Children's Dreams: Notes from the Seminar Given in 1936–1940*, L. Jung and M. Meyer-Grass (eds), Philemon Series. Princeton, New Jersey, Princeton University Press

Lindorff, D. (1995) 'Psyche, matter and synchronicity; a collaboration between C. G. Jung and Wolfgang Pauli', *Journal of Analytical Psychology*, 40: 571–86

Meier, C. A. (ed.) (2001) *Atom and Archetype; the Pauli/Jung Letters*, 1932–1958. London, Routledge; Princeton, New Jersey, New Jersey, Princeton University Press

Miller, A. (2009) *Deciphering the Cosmic Number: The Strange Friendship of Wolfgang Pauli and Carl Jung*. New York, London, W. W. Norton and Company

Yiassemides, A. (2011) Chronos in synchronicity: manifestations of the psychoid reality. *Journal of Analytical Psychology*, 56, 4: 451–71

——. (2014/2016) *Time and Timelessness: Temporality in the Theory of Carl Jung*. London and New York, Routledge

PART B

Expressions of time in the psyche

> How could something human escape the grip of time?
>
> (Scarfone 2006: 831)[1]

1 Scarfone, D. (2006). A matter of time: actual time and the production of the past. *Psychoanalytic Quarterly* 75: 807–34.

6

SCREEN TIME

Movies, mind and the experience of time

Christopher Hauke

Introduction

The poet Wislawa Symborska once commented, 'When I pronounce the word Future, the first syllable already belongs to the past.' We experience time as a one directional flow. Intuitively we divide time into past, present and future, divisions that shift instantly even as you are listening. But is this experience of time an experience of something objectively real? Relativity theory tells us that what appears as simultaneous events from one frame of reference may be seen from another perspective as two events occurring at different moments. As Einstein theorized, 'The past, present and future are only illusions, even if stubborn ones' (Davies 2012). In nature, there is no one place that we can regard as a 'moment' with a known past before it and an unknown future. For physics, time is best regarded as a block, similar to space. Like 'space' in a landscape, all of 'time' is present at once; as physicist Paul Davies says, 'Objectively, past, present and future must be equally real. All eternity is laid out in a four-dimensional block composed of time and the three spatial dimensions' (Davies 2012: 43). So maybe our experience of time as a flow is subjective: a clock does not measure the speed or flow of time; it merely measures spatially the durations between events in the similar way that a measuring tape measures distances between locations.

What gives rise to the experience of time is our *perception of change.* Change of position in space, change of light, change through deterioration or ageing (or even renewal) all lead us to conclude that one state was *earlier* than the changed state. The observation that the later state differs from the earlier one that we can recall gives us a sense of a flow of time. This, in turn, leads to the conclusion that there is a linear time running in one direction with a past, a present and a future. The past is known, the present is being known, and the future is unknown.

Connected with this perception of change is the second law of thermo-dynamics, the law of entropy, which states that matter becomes *more disorganised* over time. An intact egg has lower entropy (is less disorganised) than a smashed egg. Conforming to the second law, an intact egg can easily become a smashed egg, but a smashed egg does not reform into a whole one. Thus, 'The second law of thermodynamics plays a key role in imprinting on the world a conspicuous asymmetry between the past and future directions along the time axis' (Davies 2012: 43). Labels such as 'past' and 'future' may be applied to time directions just as 'up' and 'down' may be applied to spatial directions. But we would never talk about '*the*' up or '*the*' down; so why do we refer to '*the*' past and '*the*' future?

Going back to our unidirectionally smashed egg, if we filmed this egg breaking then ran the projection backwards, the egg would *appear to* do the impossible and reassemble itself; but this would not look real. The filming and the viewing are both experienced in time. But if we took the film and cut it up into its separate frames (twenty-four for every second of time) and stacked them up in the right order from broken egg at the top to whole egg at the bottom, 'this vertical stack retains the asymmetry implied by the arrow of time because it forms an ordered sequence in vertical space' (Davies 2012: 43). This shows that 'the asymmetry of time is actually a property of states of the world, not a property of time as such' (Ibid.). In other words, you do not need to run the film 'in time' for the arrow of time to be discerned.

It could be said that when Muybridge shot his rapid photographic sequences of humans and animals in motion he was achieving the same thing. His individual images – as David Thomson says 'ravished by the play of light on ordinary physicality and by the tiny, incremental advances through time' (Thomson 2012: 9) – revealed what had been invisible to the human eye: the succeeding states of a body in motion only previously perceived as a whole and not in its ordered parts. Celluloid movie film took up this technique in reverse running the separate images rapidly in sequence, thereby restoring the illusion of movement in the 'moving picture' or movie.

Memory and time

Memory and time are intrinsically connected aspects of how we experience ourselves and the world. The memories we hold personally, and the meaning that we construct from them, are dependent on our notion of the flow of time. Physicists realise now the concept of entropy – which leads to our experience of time flow – is closely related to the information content of a system: 'For this reason, the formation of memory is a unidirectional process – new memories add information and raise the entropy of the brain' (Davies 2012: 47). Consequently, we may register this unidirectionality as the flow of time.

Subjectively, it seems self-evident that memories are images derived from the past. But what if we viewed this the other way round? Could it be that our *experience of having memories* is the very thing that *allows the past to exist in our minds*? In other words, memories may not 'come from' the past, but the experience of a memory might be what convinces us there is 'a past' which forms part of something called time that flows in a linear fashion. Memories are regarded nowadays as a reconstructive process, the result of a creative activity rather than accurate reports of perceptions that have passed. As such, memories while appearing to be 'of the past', always include the 'present', that is the time of recall, as an unavoidable component of the phenomenon. As the idea of memory becomes less clear, our understanding of time itself follows. So how would it be if time ran differently? What if you could 'remember' what was going to happen next, in the 'future'? What would that do to your experience as a human being?

Groundhog Day

The movie *Groundhog Day* (Ramis 1993) presents the story in the straightforward style of modern Hollywood films, but the conventional narrative is compromised by the way in which it shows the main character, Phil Connors (portrayed by Bill Murray), repeating the same day's events time, after time, after time. Unlike other movies that repeat one dramatic event with different points of view or time-twists – like Kurosawa's *Rashomon* (1980) and *Run Lola Run* (1999) – *Groundhog Day* involves a low-key repetition of a single, banal day where nothing much happens. Nothing, that is, until Bill Murray's character realises that he is stuck in *repeating time*. He begins to experiment with his life, his choices and his decisions, knowing he is always destined to wake up and start the same day all over again. Watching this film, we are given an opportunity to see what life might be like having a different experience of time.[1]

Bill Murray's character is a TV weather forecaster, an arrogant man with an ego and persona-focused way of conducting his life. He is heading for a fall. He goes upstate to report on Groundhog Day, the first day of Spring, but is snowed in and must stay a second night. He wakes up to find it is Groundhog Day *again*, an awakening that repeats many times during an inexorable cycle of repeated events. Despite behaving differently himself, Phil is unable to alter or avoid what occurs during the day. Phil tries to tackle his conundrum by addressing the problem of entropy. He breaks a pencil and leaves it on the night stand; the law of entropy says it cannot go back to its organised state. Or can it?

After some time, Phil is able to let go of his one-sided ego-consciousness and, having realised the tragic hopelessness of his position, he decides to experiment with attempts to kill himself. Only at this moment, which signifies the death for his former personality, does something get reformed. As Jung says, 'It is as though, at the climax of the illness, the destructive powers were converted into healing forces' (Jung 1932: para. 534). In this case, the weatherman's reconnection with the neglected aspects of himself is signalled by his attempts to make true connections with others.

What can we learn from this story about time? The story of Phil the weatherman reminds me of two aspects of psychological suffering and the opportunity offered by psychotherapy. First, we could see him like someone who – through fear, anxiety and insecurity – clings to known habits and obsessions, including a rigid and narrow sense of his own identity. Through this approach, he achieves a form of self-containment which protects his vulnerability; but, like the Uroboros, this closed circle of containment prevents further inner and outer connections which could ameliorate his position.

Second, that the Groundhog Day itself with its own circular Uroboric time *is imposed* on the weatherman; in fact, it provides the very opportunity that this soul needs. By preventing him from moving on in time, the endless Groundhog Day forces him to reflect on his own attitude. At the same time, it provides a containment – an endless, timeless opportunity for his soul to access this insight.

We can see the similarity between the containing Day and the containment of psychotherapy. Both offer an opportunity for the dominant conscious attitude to be challenged. At the same time, both entail suffering: the suffering connected with the first insight into the self and the self's possibilities, which are not yet available. This moment, which seems like a beginning is, in fact, also an ending; it is the death of the old personality structures. It is the beginning of the ending, since it does not even allow for giving up. In the film, Phil commits suicide several times but, of course, wakes up every day in the same hotel room on the same Groundhog Day. Kafka expressed this fear of the Herculean task of addressing the unconscious: 'A first sign of nascent knowledge is the desire for death. This life seems unendurable, any other unattainable' (Kafka 1970: 164).

Movies and time

Could any other creative art form tell this story as well as film does? Movies are experienced in linear time which is dictated by the form of film itself. The same holds true for theatre, music and dance. But unlike these art forms, a film never gives a varied performance; music, theatre and dance can be interpreted and performed faster or slower on different occasions. Movies cannot do that; a film always runs for a fixed time (even when you get a new cut of the film, since that as well is fixed). Hence, all other time-based art forms differ from film. In painting, sculpture, photography, poetry and literature time is not intrinsically linked to, or embedded in, the art object itself: The length of time of the viewing experience is governed by the viewer or reader and how long they want to spend with the object, and not by the art form itself.[2]

Movies are experienced as a linear flow of time that accompanies the viewer's own lived time. After watching *Groundhog Day*, you will be 97 minutes older. But nothing will have changed in the film itself. Watching it today, the actor Bill Murray looks considerably younger than he does now, in 2017, since the film was made in 1993. In fact, in twenty years he will not have aged, *but* every viewer will have! A painting or sculpture shares the same time as its viewers and will age or deteriorate across the same time span, whether they are present or not. All are subject to the law of entropy. Of course, the physical matter of the film or DVD disc is also deteriorating, but the *filmed content of the movie is fixed forever in time.* Which, when you come to think of it, is a bit like the Groundhog Day itself.

Movies, emotion and time

Movies are primarily an experience of emotion – as opposed to cognition – so perhaps the manipulation of time is an important contribution to such emotional experiences. Different viewings of the same film, in different circumstances, may affect the subjective sense of time passing. The director Anthony Minghella once remarked that although the film lasts a fixed period of time, 'If Walter Murch [his editor] and I are watching, it goes at one speed . . . If there are ten people, a hundred people, the collective weight on the brake of our car is extraordinary. If everybody walks away and I watch it by myself, I find it goes very quickly. The point is the film changes in front of you' (Koppelman 2005: 163–64).

Uri Hasson, a Princeton psychologist, has found that activity in the same areas of the brain – in the primary auditory and visual cortex – as well as the fusiform face area rise and fall synchronously when a group of viewers watch the same movie. Moreover, structured movies with cuts and angles have this effect to a larger extend than unstructured movies. These scans show the strongly synced effects when watching *The Good, the Bad and the Ugly* and *Dog Day Afternoon*; there are only weak effects for the less structured, improvised show *Curb Your Enthusiasm* and even less for a non-dramatic reality film clip (Miller 2014).

When considering our experience of time while watching a movie, the psychiatrist Antonio Damasio discusses the case of Alfred Hitchcock's film *Rope* (Hitchcock 1948) which was filmed with an unusual technique. Instead of scenes being shot (sometimes with several takes) then edited, Hitchcock rehearsed the actors and the camera moves, and then shot the film in one continuous take. As a canister only holds ten minutes of film, he originally shot ten reels of ten minutes each, using a live technique to mask the cross-over between each long take.

The narrative story involves a murdered body hidden in a flat during a dinner party. Hitchcock states that the story runs for 105 minutes, from 7.30 to 9.15 in the evening (Truffaut 1984: 179). Through the window, the light of New York City's sky-line changes very gradually to imitate the setting sun. However (without the credits), the filmed action itself is only 72 minutes long. Damasio asks: 'What has happened to those missing minutes?' We do not experience the film as shorter than the 105 minutes; some might even perceive it as longer!

Damasio suggest three main reasons. The first is the way that the artificially created light accurately imitates the setting sun's movement across the sky and the beginning of night-time. The rhythm of day and night is vital to our inner clock, and the film uses this to convince us of the reality of its story-time. Second, the period of eating the food at the party before it is all cleared away goes rapidly in real time (about 20 minutes) but we can see the guests eating in a relaxed manner, therefore we attribute a longer time to that event. Lastly, Damasio cites the techniques of film itself: in the language of film, transitions (known as a 'dissolve', or 'fade-to-black') signal a passage of time. Taking only a fraction of a second each time, there are several of these during the film, a factor which must contribute to the sense of a longer period of time passing.

Damasio's lab experiments 'suggest that the brain generates images at faster rates when we are experiencing positive emotions . . . and reduces the rate of image-making during negative emotions' (Damasio 2012: Loc 701). For the film, it again shows how 'the experience of duration is a construct . . . based on . . . the content of events being perceived, the emotional reactions those events provoke and the way the images are presented to us, as well as the conscious and unconscious inferences that accompany the film's images' (Ibid.).

Depth psychology, memory and time

Time is a way of organising conscious awareness, memories and human activity. The psychoanalytic concept of the unconscious is predicated on experience, time, memory, repression, projection and recollection, and their emotional effects on the mind. At the beginning of the twentieth century, at the same time that psychoanalysis and movies were gaining popularity, the psychologist William James questioned the concept of 'a moment of time'. He concluded that the present moment was not so much a 'knife-edge' but more a 'saddleback, with a certain breadth of its own on which we sit perched and from which we look in two directions into time' (Gifford 1990: 102).

What happens when conscious memory fails or is limited to only the past fifteen minutes, as it happens with the character in Christopher Nolan's film *Memento* (Nolan 2000)? Made in 2000 (exactly a hundred years after Freud's *The Interpretation of Dreams*), Nolan's protagonist uses several techniques as surrogates for his lack of memory: He takes instant Polaroid photographs and writes messages to himself on them. But his ultimate technique, so that nothing may get lost, is to tattoo essential information on his own body. His 'memory' is externalised outside his mind, and embodied. This reminds me of the prisoners in Kafka's story *In the Penal Colony* who had descriptions of their crimes permanently etched in words onto their own skin to prevent them from even trying to forget why they were locked up.

Conclusion: the healing moment of time

It is the metaphor of the mechanisation of time that persuades us of the fine exactitude of a moment in time that the photograph has supposedly captured. I think we can push this metaphor back on itself for our purposes. Picture this!

What if we step back for a moment, and regard our everyday consciousness as equivalent to a photograph? Or, more accurately, let us liken the activity of our conscious mind to an endless sequence of photographs, like the images flowing at 24 frames-per-second, which produce the illusion of the movie picture. This is *Kairos* – the cleverness of conscious time. But the ground of all this, *the rich unfrozen movement* from which all these 'still' frames derive, is the unconscious. The unconscious can be likened to the other 'real' life being lived outside and beyond the lens of consciousness, sometimes coinciding, but ultimately being independent of it.

Jung points out how modern consciousness has lost certain functions that were once experienced consciously by humans. In the present day, with the dominance of rational thinking, certain symbolic functions of the mind were driven down, to form the contents of the unconscious. Jung's version of remembrance, that is, bringing the past into the present, involves the archetypes and our symbolic life far more than our childhood and family dynamics.

Notes

1 Even more so than in other films like Tom Tykwer's 1999 *Run Lola Run*, Frank Capra's *It's a Wonderful Life* and Kurosawa's *Rashomon* where events and people re-encountered, in the course of the hero's journey toward self-knowledge.
2 Incidentally, I am noticing how I am using spatial concepts such as 'length' and 'long' or 'short' to describe this experience of time, rather reinforcing its intrinsic spatiality.

References

Damasio, A. (2012). Remembering when. *Scientific American* 21(1): 565–715, in Amazon Kindle Version Collection: *Scientific American. A Question of Time*

Davies, P. (2012). That mysterious flow. *Scientific American* 21(1): 165–294

Gifford, D. (1990). *The Farther Shore; A Natural History of Perception 1798–1984*, London, Faber

Groundhog Day. (1993). H. Ramis (dir.), T. Albert (prod.), USA, Columbia Pictures

It's a Wonderful Life. (1946). F. Capra (dir. and prod.), USA, Liberty Films

Jung, C. G. (1932). *Psychotherapists or the Clergy*. In C. W. Vol. 11, London, Routledge & Kegan Paul

Kafka, F. (1970). Reflections on sin, pain, hope and the true way, in *The Great Wall of China: Stories and Reflections*, New York, Schocken Books

Koppelman, C. (2005). *Behind the Seen. How Walter Murch edited 'Cold Mountain' using Apple's Final Cut Pro and what this Means for Cinema*, Berkeley, California, New Riders

Memento. (2000). C. Nolan (dir.), C. Ball and W. Tyrer (prod.), USA, Newmarket Films, Summit Entertainment

Miller, G. (2014). www.wired.com/2014/08/cinema-science-mind-meld/

Rashomon. (1950). A. Kurosawa (dir.), M. Jingo and M. Nagata (prod.), Japan, Daiei Motion Picture Company

Rope. (1948). A. Hitchcock (dir.), S. Bernstein and A. Hitchcock (prod.), USA, Warner Bros., Transatlantic Pictures, MGM

Run Lola Run. (1999). T. Twyker (dir.) S. Arndt and G. Henke (prod.), Germany, X-Filme Creative Pool, Westdeutscher Rundfunk

Thomson, D. (2012). *The Big Screen; The Story of the Movies and What They Did to Us*, London, Penguin

Truffaut, F. (1984). (With the collaboration of Helen G. Scott), *Hitchcock*, New York and London, Simon and Schuster

7

JUST A MOMENT? OR, THE ARCHAIC AS AN EXPRESSION OF THE ETERNAL IN TIME

Paul Bishop

Towards the end of the first episode of the three-part documentary series *Magnetic North*, its presenter, the writer and cultural critic, Jonathan Meades, visits the famous chalk cliffs on the island of Rügen. Famous not least, that is, because of their depiction in a famous painting by the German Romantic artist Caspar David Friedrich (1774–1840), his *Chalk Cliffs on Rügen* (1818–1820).[1] Conventionally, this picture is read as representing an Enlightenment, or even a revolutionary, message: its three figures are looking out towards a bright new dawn.[2] As one commentator on the painting has written: 'The crystalline, deathly rigidity of the rocks is overcome by the sea and, ultimately, by the spirituality achieved in the sky. Matter is dissolved in a progression towards depth. Two sailing boats, probably to be interpreted as ships of life, move into the distance, while a third, barely visible, is disappearing across the horizon. In the sky, veiled by its branches, the picture finds its culmination' (Wegmann 1993: 84). But instead, Meades suggests, the three figures depicted are not so much gazing out to sea, as turning away from the immensity of the forest behind them: 'What Friedrich's painting does not reveal is the forest behind him, behind the figures staring transfixedly at the sea and at a limitless world.' As Meades goes on to explain: 'For Friedrich, as for the Brothers Grimm, forests were places of real and metaphorical darkness which mankind had ceased to dwell in and worship in; which mankind had escaped from, but which still exerted a morbid attraction.'

Meades's suggestion is a compelling one, and it prompts us to ask why these figures should be turning away from the forest. To this question Meades has already given one possible answer, in his portrayal of the forest as a location of the primitive, the primordial, or the archaic. 'The fear that overtook me,' he says, 'was primal; it went beyond that German conditioning, even though I was in

Germany. Brain and body were united in foreboding; it was sufficient to persuade me that another memory might exist – an atavistic memory. I was the distant legatee of some distant, woad-smeared forebear's fear of the forest.' At this point, Meades turns directly to the camera and tells us: 'In places like this, reason does not sleep: it is concussed by a heavy blow. So I wouldn't have been surprised to see berserker warriors, in bloody furs and skin, or knots in wood mutating into witches' faces, or fabulous stags with shining bez tines, or enchanting wood sprites whose kiss is fatal. This place seemed to contain,' he emphasizes, 'an inventory of horrible possibilities.'

In typical Meadesian fashion, this commentary is at once facetious and profound. And what Meades has laid his finger on here is something essential about the archetypal dimension of time – recall his remark about an 'atavistic memory', which sounds almost Jungian – and its relation to the psyche. (Indeed, it is to Jung that we are indebted for an entire working-out of the notion of the archaic, in a talk given in October 1930 and subsequently published as 'Archaic Man'.[3]) For what characterizes localized depictions of the archaic as one finds it in Friedrich is precisely this emphasis on a particular dimension of time. And one finds it in many other works too, some of which choose to focus not on cliffs – but on caves.

After all, as the Traditionalist thinker René Guénon (1886–1951) pointed out in his *Symboles fondamentaux de la Science sacrée*, caves – like mountains – represent forms of the cosmos, symbolizing 'spiritual centres' (Guénon 1995: 139–71).[4] (For instance, the 'cave of the heart' is a well-known traditional expression: in Sanskrit, *guhā* denotes a cave but is also applied to the inward cavity of the heart; it comes from the root, *guh*, meaning 'covering' or 'hiding', as does a similar root, *gup*, from which *gupta*, applied to everything secret, derives, and which is the equivalent of the Greek *kruptos*, the root of 'crypt' as a synonym of 'cave'). While Guénon takes as his starting-point a study by the classicist and Virgil scholar, W. F. Jackson Knight (1895–1964), of the description of the gates to the cave of the Cumean Sibyl in the sixth book of the *Aeneid* (Knight 1936), there are other many possible examples to be found across the centuries – back to and including Palaeolithic cave art which has been said to offer 'an intuition of eternity'.[5]

Here, we shall confine ourselves to some well-known literary examples. For instance, in a passage in his *Metamorphoses*, Ovid (43 BCE–17 CE) describes a physical landscape in terms that evoke its timeless and its numinous (and, in this sense, its archaic) atmosphere:

> There was a valley clothed in hanging woods
> Of pine and cypress, named Gargaphie,
> Sacred to chaste Diana, huntress queen.
> Deep in its farthest combe, framed by the woods,
> A cave lay hid, not fashioned by man's art,

> But nature's talent copied artistry,
> For in the living limestone she had carved
> A natural arch; and there a limpid spring
> Flowed lightly babbling with a grassy sward.
> Here, tired after the hunt, the goddess loved
> Her nymphs to bathe her with the water's balm.
>
> (Ovid 1987: 55–56)

Then again, in one of the leading representatives of late Stoicism, the philosopher Seneca (c. 4 BCE–65 CE), who lived on the very cusp of the pagan and the Christian eras, sought to persuade his correspondent, Lucilius, of the existence of divinity by evoking the sense of its archaic – in the sense of 'originary', 'primordial', or again, 'numinous' – presence in nature:

> If you're confronted by some dense grove of aged and giant trees shutting out every glimpse of sky with screen upon screen of branches, the towering stems, the solitude, the sense of strangeness in a dusk so deep and unbroken where no roof is, will make deity real to you. Again, the cavern that holds a hill-side poised on its deep-tunnelled galleries of rock, hollowed into that roomy vastness by nature's tools, not man's, will strike some hint of sanctity into your soul. We render homage to the sources of great rivers. The vast stream that bursts suddenly from hiding has its altars; hot springs are worshipped; there are lakes hallowed by their dark inscrutable waters and unplumbed depths.
>
> (Seneca 1932: 127–28)

A couple of centuries later, in his celebrated essay 'On the Cave of the Nymphs' the Neoplatonic philosopher Porphyry (c. 234–305) went back to Homer's account of a cave. In book 13 of the *Odyssey*, one finds the following description of the cave in which Odysseus hides the gifts of King Alcinoos of the Phaeacians:

> At the head of the harbour is an olive with long leaves,
> And close to that is a pleasant and shadowy cavern
> Sacred to the nymphs who are called Naiades.
> And in it there are mixing bowls and two-handled jars
> Of stone. And the bees store up the honey in them.
> There are very long stone looms in it, where the nymphs
> Weave sea purple mantles, a wonder to behold,
> And ever-flowing waters are there. It has two doors,
> One toward the North Wind, accessible to men,
> And the other one divine is toward the South Wind, nor may men
> Approach by that one, but it is a path for immortals.
>
> (Homer 1993: book 13, ll. 102–12)

In an astonishing interpretative *tour de force*, Porphyry uncovers all that is universal – indeed, archetypal – about this literary cave. For Porphyry, the ancient custom of dedicating caves and grottoes to the universe contained an important truth, teaching us 'to see the earth as a symbol of the matter from which the universe is constructed' (Porphyry 1999: 141).[6] For Porphyry, 'the universe is a natural construction of matter' and the stone and rock of the cave are 'symbolic of matter's inert nature and resistance to form, and its unboundedness of matter's shapelessness' (Porphyry 1999: 141). On Porphyry's reading, every detail in Homer's description counts: the olive tree is sacred to Athene, the goddess of wisdom; its wetness and humidity, or its 'dark and "shadowy" appearance' symbolize the 'material character' of the universe; the stone bowls and storage vessels represent souls descending to birth in the body; the garments of sea purple stand for flesh formed from blood, symbolizing the body as a tunic in which the soul is clothed; the nymphs are like bees, because they give delight; honey has powers of purification and preservation; birth is assisted by water, like the spring waters; finally, the two entrances to the cave (mirroring the constellations of Cancer and Capricorn) serve as the gate through which souls descend and the gate through which they ascend again to the gods. In short, not only is the universe 'governed by Intellect and an eternal and everlasting Wisdom' but such previous interpreters as Numenius were right to see Odysseus as 'a symbol of the man who passes through the temporal cycle, and is thus restored to that company who live beyond the waves of the sea of time' (Porphyry 1999: 145). What this cave represents is, in a sense, timeless.

For some writers and thinkers there is something special about caves; for others, it is the forest (even if, as in Friedrich's Rügen painting, it is occluded). Moving forward in time, we find in a famous essay, *Nature*, by the American Transcendental philosopher Ralph Waldo Emerson (1803–1882), a remarkable passage in praise of the kind of experience he associated with the forest or the woods:

> Crossing a bare common, in snow puddles, at twilight, under a clouded sky, without having in my thoughts any occurrence of special good fortune, I have enjoyed a perfect exhilaration. I am glad to the brink of fear. In the woods, too, a man casts off his years, as the snake his slough, at what period so ever of life, is always a child. In the woods, is perpetual youth. Within these plantations of God, a decorum and sanctity reign, a perennial festival is dressed, and the guest sees not how he should tire of them in a thousand years. In the woods, we return to reason and faith. There I feel nothing can befall me in life – no disgrace, no calamity, (leaving me my eyes), which nature cannot repair. Standing on the bare ground – my head bathed by the blithe air, and uplifted into infinite space – all mean egotism vanishes. I am become a transparent eye-ball; I am nothing; I see all; the currents of the Universal Being circulate through me; I am part or particle of God.
>
> (Emerson 1983: 10)

The impulse to return to the forest was equally well understood by another transcendentalist philosopher, Henry David Thoreau (1817–1862), who took the decision in 1845 to go and live in a small, self-constructed cabin on the shores of Walden Pond in Concord, Massachusetts. Thoreau used his reflections during his two years, two months, and two days living here as the basis for his book, *Walden; or, a Life in the Woods* (1854). Thoreau tells us: 'I went to the woods because I wished to live deliberately [. . .] I wanted to live deep and suck out all the marrow of life' (Thoreau 1908: 78–79). A good example of the kind of heightened phenomenological awareness to which such 'deliberateness' can give rise can be found in Thoreau's address to the Middlesex (Mass.) Agricultural Society under the title *The Succession of Forest Trees* (Thoreau 2001: 429–43).

Moreover, this return to the forest recalls the ancient legend according to which the philosopher Democritus, returning to Greece from a long voyage to India, decides to build a hut at the end of his garden and to spend the remainder of his life there. This legend, as well as the experiences of Emerson and Thoreau, inspired the contemporary French philosopher Michel Onfray to write a poetic text, *Le Recours aux forêts: La tentation de Démocrite* (2009), subsequently staged and choreographed by Jean Lambert-Wild and Carolyn Carlson and performed by the Comédie de Caen in 2011. After the first part, where Democritus laments what he has learnt about humanity on his travels, the second part consists of a series of intense lyrical exclamations:

> Above my head, often, I know heaven to be empty of God / but full of stars / And, among them all, a single one, a unique one / The indestructible magnetism of a compass for Being.[7]
>
> (Onfray 2009: 56; my translation)

But what Emerson, Thoreau and Onfray praise with such innocence – 'In the woods, we return to reason and faith' – becomes more ideologically complicated, even suspect, in the case of a film called *Ewiger Wald*, and sometimes known by its international English title, *Enchanted Forest.* First shown in Munich in 1936, this film was directed by Hanns Springer and Rolf von Sonjevski-Jamrowski; but it had been commissioned in 1934 by the Militant League for German Culture, a National Socialist cultural organization headed up by Alfred Rosenberg. What in the case of Emerson is transcendental idealism, strays close to – and over the line into – pure ideology. For here the forest that sheltered the ancient tribes of Germania, gave protection to Arminius (the chieftain of the Germanic Cherusci who defeated the Romans in a battle in the Teutoburg Forest in 9 CE), and provided sanctuary for the Teutonic Order of knights, is presented as standing in the face of the Great Peasants' Revolt of 1524–1525, as being destroyed by the forces of industry, and being subject to humiliation by the soldiers of the French occupying army of 1918, among them soldiers of black ethnicity. When one watches this film, a very clear political message emerges: The forest is German, it belongs to the Germans; far from being 'eternal', *this* forest is very clearly anchored in the political ideology of the Third Reich.[8] (For this reason, the present writer feels it is easier to talk about the archaic in English to UK or US audiences than to talk about it in German to German or Swiss audiences, where the entire language of the archaic has become contaminated.)

Yet what Ovid, Seneca, Porphyry, Walden and Thoreau, and in a way even Democritus and Onfray, are talking about is a sense of what the German theologian Rudolf Otto (1869–1937) once formulated as 'the sacred' (*das Heilige*); that is, a sense of 'the holy' with its different aspects of the numinous, of *mysterium tremendum*, and of *fascinosum* (Otto 1958). This response is what, in the Mothers Scene of *Faust*, Part Two, is described as a 'shudder' (*Schaudern*)[9] (Goethe 2001: 178); it is an expression of something which, in the view of the French intellectual historian, Pierre Hadot (1922–2010), constitutes the expression of specifically modern sensibility, one found subsequently in F. W. J. von Schelling, Friedrich Nietzsche, Rainer Maria Rilke and Martin Heidegger (Hadot 2001: 172) even if this *frisson*, this *nuance d'angoisse*, is clearly something that goes much further back. In fact, in an interview Hadot once recalled two life-changing experiences, one beneath a star-lit sky, the other inside a room at home, that demonstrates the persistence of this kind of experience in the modern (or even postmodern) world:

> I was filled with an anxiety that was both terrifying and delicious, provoked by the sentiment of the presence of the world, or of the Whole, and of me in that world. In fact, I was incapable of formulating my experience, but after the fact I felt that it might correspond to questions such as *What am I? Why am I here? What is this world I am in?* I experienced a sentiment of strangeness, of astonishment, and of wonder at being there. At the same time I had the sentiment of being immersed in the world, of being a part of it, the world extending from the smallest blade of grass to the stars. This world was present to me, immensely present. Much later I would discover that this awareness of belonging to the Whole was what Romain Rolland called the 'oceanic sentiment.' I believe that I have been a philosopher since that time, if by philosophy one means this awareness of existence, of being-in-the-world.
>
> (Hadot 2011: 5)

So why should forests and caves matter so much? What is the significance of these natural locations for psychoanalysis? To understand why they matter, and what their symbolic significance is, we could turn to C. G. Jung. Or we could turn to one of his ideological opponents, the German sociologist and member of the Frankfurt School, Theodor W. Adorno (1903–1969).

In his *Dialectic of Enlightenment* (1944), co-authored with Max Horkheimer (1895–1973), Adorno developed a profound critique of modernity. Far from being opposites, he argued, mythology and science or myth and Enlightenment were dialectically interrelated, each presupposing the other. At a crucial point in his argument, Adorno speaks of the 'remembrance of nature in the subject' as being the central point at which the Enlightenment is opposed to tyranny. 'By virtue of this remembrance of nature in the subject,' he and Horkheimer wrote, 'in whose fulfilment the unacknowledged truth of all culture lies hidden, enlightenment is universally opposed to domination' (Horkheimer and Adorno 1996: 40). Now this phrase and its idea, one of an 'inner nature', enjoy a distinguished and significant history. In fact, it had originally been formulated by Sigmund Freud, who speaks in *Civilization and its Discontents* (1930) about how 'a piece of unconquerable nature' forms part of our psychic constitution (Freud [1930] 1961: 86). Thereafter, the notion of 'inner nature' crops up, time and again, in the thinking of the first and second generations of the Frankfurt School.

In his *Theory of Communicative Action* (1981), for example, Jürgen Habermas (b. 1929) focuses on this key phrase. As part of *his* critique of Adorno's and Horkheimer's critique of instrumental reason, Habermas accuses his colleagues of 'follow[ing] the (largely effaced) path that leads back to the origins of instrumental reason, so as to *outdo* the concept of objective reason'; thus their theory of mimesis, Habermas argues, leads them to speak about it 'only as they would about a piece of uncomprehended nature'. Habermas's judgment on Adorno is particularly severe: 'As opposed as the intentions behind their respective philosophies of history are, Adorno is in the end very similar to Martin Heidegger (1889–1976) as regards his position on the theoretical claims of objectivating thought and of reflection: the mindfulness [*Eingedenken*] of nature comes shockingly close to the recollection [*Andenken*] of being' (Habermas 1984/1987).[10] Of course, from another point of view, the shared legacy with Heidegger could alternatively be seen as a strength, not a weakness, and as evidence of the global appeal and universal truth of this topos.[11]

However, although the expression 'inner nature' is always attributed to Freud, the phrase is originally found in C.G. Jung, who writes in his paper *Psychological Types* (1923) that 'the unconscious is the residue of unconquered nature in us, just as it is also the matrix of our unborn future' (Jung 1923: para. 971). And in his later paper *Analytische Psychologie und Weltanschauung* (1927/1931), Jung remarks that 'the fantasy-images of the unconscious which our rationalism has rejected' constitute 'part of the nature *in us*' (Jung 1927/1931: para. 739). Indeed, the insistence on the persistence of the archaic – the sense that, like the poor in the gospel, the primordial is always with us – belongs to the major hallmarks of Jungian thought, and the notion of 'inner nature' represents an important point of intersection between analytical psychology and critical theory (Evers 1987: 186–87).

Nietzsche, Zarathustra and *Augenblick*

Given (a) the natural symbolism of the cave and the forest as representing the persistence of the archaic; and (b) the idea of an 'inner nature' within each individual, I would like, using a Jungian interpretative apparatus, to turn to one of the most remarkable texts ever written, *Thus Spoke Zarathustra* by Friedrich Nietzsche (1844–1900). (This is the sort of reading Jung himself attempted in his famous seminar on *Zarathustra* from 1934 to 1939 [Jung 1988]). Far from being an abstract philosophical text, this is a work full of natural imagery which cries out for an allegorical reading.

At the beginning of this work, Zarathustra – already living in Walden-like circumstances in his home by a lake – intensifies his solitude even further by going 'into the mountains' and living for ten years in a cave. The text is full of references to forests, mountains and islands; to the sun, clouds and lightning; to trees, birds and animals. Sometimes the references to nature appear whimsical, as in the beginning of the chapter in Part Two entitled 'Of Scholars':

> When I lay asleep, then did a sheep eat at the ivy-wreath on my head, it ate, and said thereby: 'Zarathustra is no longer a scholar.' / It said this, and went away clumsily and proudly. A child told it to me. / I like to lie here where the children play, beside the ruined wall, among thistles and red poppies.
>
> (Nietzsche 1917: 135–36)

Then again, in 'Of the Priests', Zarathustra declares: 'Only when the clear sky again looks through broken roofs and down upon grass and red poppies on broken walls – only then will I turn my heart again towards the places of this God'[12] (Nietzsche 1969: 115). Sometimes, however, the natural setting is of profound significance, such as the case with a chapter in Part Four entitled 'At Noontide'.

Because Jung's seminar on *Zarathustra* never got as far as discussing this chapter (or any other chapter in Part Four), we must look elsewhere for his comments on 'At Noontide'. We find them in a paper delivered at the Eranos Conference of 1939, dedicated to the various aspects of rebirth (Jung 1940/1950: para. 199–258). According to Jung, what Nietzsche had experienced (along with analogous experiences from various cultures) constituted a primordial affirmation of humankind, an instance of what Jung defined – to use his scintillating expression that still causes much conceptual vexation – an archetype (Jung, 1940/1950: para. 207). 'The archetype of rebirth' is an unusual one, being based on an *experience* rather than on a figure, either abstract (the Self) or concrete (the Mother, the Wise Old Man, the anima, the animus, or the Child). Yet, as with all such archetypes, Jung was able to cite numerous examples of its manifestation across different cultures, including the classical, Mithraic, Qu'ranic and contemporary European ones.

Jung distinguished between two kinds of archetypal rebirth experience: One of the transcendence of life in general (subdivided into ritual-induced and immediate experiences), the other of one's own subjective transformation (divided into a further eight sub-categories). For Jung, the midday vision experienced by Zarathustra in 'At Noontide' belongs in the former category of the transcendence of life, and constitutes a classic example of the spontaneous, ecstatic, or visionary experience. Indeed, Jung considered its details to be so accurate that he wrote 'it is just as if Nietzsche had been present' at a performance of the Mysteries of Eleusis, the ancient initiation ceremonies held annually at Eleusis, just outside Athens, in honour of Demeter and Persephone (Jung 1940/1950: para 210).[13] According to the Platonic and Neoplatonic translator and commentator Thomas Taylor (1758–1835), such greater mysteries as the Eleusinian Mysteries 'obscurely intimated, by mystic and splendid visions, the felicity of the soul both here and hereafter, when purified from the defilements of a material nature, and constantly elevated to the realities of intellectual vision' (Taylor 1995: 77). Or, in Platonic terms, 'the ultimate design of the mysteries' was 'to lead us back to the principles from which we descended, that is, to a perfect enjoyment of intellectual good'; the tradition of these principles constituted 'one part of the doctrine contained in the *απόρρητα* or secret discourses', and 'the different purifications exhibited in these rites, in conjunction with initiation and inspection, were symbols of the gradation of virtues requisite to this reascent of the soul' (Taylor 1995: 64, 115).[14] In the memorable phrase from Henry Wadsworth Longfellow's poem *A Psalm of Life*, 'The soul is dead that slumbers'; and according to the Canadian-born esoteric philosopher Manly P. Hall (1901–1990), this line 'strikes the keynote of the Eleusinian Mysteries' (Hall 2003).

In one sense, Jung is entirely correct to associate Nietzsche with the great Mysteries: as Socrates remarks in the *Phaedo*, 'bear the emblems, but the devotees are few', and 'these devotees are simply those who have lived the philosophical life in the right way' (Plato 1989: 52). Moreover, Nietzsche himself attached great importance to the ancient Mysteries, specifically the Dionysian ones.[15] In his late work, *Twilight of the Idols* (1888/1968), he argued, just as he had in his earlier (and unpublished) sketch for a treatise entitled *Origin and Goal of Tragedy* (1870/1871), that without a knowledge of the ancient mysteries, Greek art in general, and tragedy in particular, would remain incomprehensible. 'For it is only in the Dionysian mysteries, in the psychology of the Dionysian condition, that the *fundamental fact* of the Hellenic instinct expresses itself – its "will to life,"' Nietzsche argued, asking: '*What* did the Hellene guarantee to himself with these mysteries?', and answering immediately: '*Eternal* life, the eternal recurrence of life; the future promised and consecrated in the past' (Nietzsche 1888/1968: 109).

Jung is completely right in another sense, too. It *is* 'just as if Nietzsche had been 'at the mysteries of Eleusis, for in the chapter entitled 'At Noontide' all the classical *topoi* of the hour of midday in antiquity – or the hour of Pan – are present,' and Zarathustra accedes to an archaic dimension of time. The sun stands over

Zarathustra's head; the landscape is bucolic, pastoral; from the vine, yellow grapes offer themselves to the wanderer (Nietzsche 1969: 286–87). In this rich and vivid description, one detail stands out: 'the old gnarled and crooked tree'. Surely this tree – which reminds us of the solitary trees standing in the midst of devastated, wind-blasted landscapes in works by Caspar David Friedrich – is meant to recall Zarathustra's ambition in 'Of Involuntary Bliss', when he was four days' journey away from the blissful islands, that in the 'afternoon of [his] life' he will uproot the trees of this blissful islands and set up each one to stand by the sea, 'gnarled and twisted' yet with a 'supple hardiness' to serve as 'a living lighthouse of unconquerable life'?[16] (Nietzsche 1969: 181–82).

Because of the heat at noonday, Zarathustra begins to feel thirsty and he wishes to refresh himself from the grapes, reaching out his arm to pluck one; as he does so, he succumbs to an even greater wish: to fall asleep. And so he does (Nietzsche 1969: 287)[17]. As Zarathustra falls asleep, his eyes stay open. Is what he then experiences a vision? A dream? A moment of ecstasy? An out-of-the-body experience? Or a truly-and-profoundly-bodily experience? For the German phenomenologist and vitalist philosopher Ludwig Klages (1872–1956), the essence of the Dionysian mysteries lies in the experience of ecstasy: 'Since the time of Nietzsche our knowledge of these matters' – i.e., the 'holy mysteries, which the dense shade of horrific myths hid from the eye of the uninitiated' – 'is chiefly connected with the name of the Thraco-Grecian god Dionysus, who in truth was only one manifestation of this universal elementary power with certain peculiarities of partly barbarian and partly Hellenic origin.' Klages goes on to explain:

> Our views of the psychical conditions of his 'epiphany' are confirmed by the fact that the mystic took the perfected degree to be not an 'exaltation' or 'purification' of the soul, but simply an 'ecstasy', to be taken literally as 'being beside himself', and that he prepared for it by a profound excitement of the senses, whose deadly was softened only by the 'rage' of orgiastic dances. But, as being beside himself, the person who has been disassociated from himself has become 'enthousiasmos', that is, 'filled by the god' or 'possessed.' He sees no longer with the eyes of common day, bounded by space, but is beyond every barrier, even that of time; words which he utters have prophetic meaning, and his power is capable of magical effect at a distance. And when enthusiasm fertilizes knowledge, then, according to the belief of every age and race, there arises 'revelation', 'illumination', 'enlightenment', 'inspiration.'
>
> (Klages 1929: 219)

As in the moment depicted in *Human, All Too Human*, 'only [Zarathustra's] eyes are alive' and the experience that follows could well be described as 'a death with open eyes'. (According to Sheridan Hough, 'anyone surging toward a noontide epiphany runs the considerable risk of being lost in that rapture forever' and this is why noontide is 'a death with open eyes' – and an abyss or *Abgrund*) (Hough 1997: 109). As he falls asleep Zarathustra speaks – and not for the first time, nor for the last – to his heart.

Thus begins Zarathustra's monologue – or, more precisely his dialogue with his heart or with his soul. Later, in the course of his monumental *Red Book*, Jung will similarly address his soul in this way; in a chapter in *Liber primus* entitled 'Refinding the Soul', Jung cries out: 'My soul, where are you? Do you hear me? I speak, I call you – are you there? I have returned, I am here again. I have shaken the dust of all the lands from my feet, and I have come to you, I am with you. After long years of long wandering, I have come to you again' (Jung 2009: 232). In this respect, as in so many others, *Zarathustra* provides Jung with a textual basis for his exploration of the psyche in the *Red Book*, and it is not for nothing that Wolfgang Giegerich has described the *Red Book* as Jung's answer to Nietzsche (Giegerich 2010: 376).[18]

Halfway through 'At Noontide' the bucolic or idyllic aspects of this chapter reach their intensity, alluding to the special quality of the noontide hour, of midday, as the hour of Pan.[19] Amid its intense happiness (*O Glück! O Glück!*), Zarathustra asks his soul if it wishes to sing (Nietzsche 1969: 288).[20] For singing is a sign of convalescence, Zarathustra's animals have told us,[21] and earlier, in 'Of the Great Longing', Zarathustra had bidden his soul sing:

> O my soul, now have I given thee all, and even my last possession, and all my hands have become empty by thee: *that I bade thee sing*, behold, that was my last thing to give! / That I bade thee sing, say now, say: *which* of us now oweth thanks? Better still, however: sing unto me, sing, O my soul! And let me thank thee![22]
>
> (Nietzsche 1917: 251)

In his *Attempt at a Self-Criticism* appended to the second edition of *The Birth of Tragedy* Nietzsche would write: 'What spoke here – as was admitted, not without suspicion – was something like a mystical, almost maenadic soul that stammered with difficulty, a feat of the will, as in a strange tongue, almost undecided whether it should communicate itself or conceal itself. It should have *sung*, this "new soul" – and not spoken!' (Nietzsche 1968: 20). Seen in this light, *Thus Spoke Zarathustra* itself is this song of the new soul, and it is significant that the German Symbolist poet Stefan George (1868–1933) alludes to this passage in his 'Nietzsche' poem in *The Seventh Ring* (1907).[23]

'If ever I spread out a still sky above myself and flew with my own wings into my own sky,' Zarathustra had sung in 'The Seven Seals' at the end of Part Three, 'if, playing, I have swum into deep light-distances' – which is exactly the situation in 'At Noontide' – 'thus speaks bird-wisdom: "Behold, there is no above, no below! Fling yourself about, out, back, weightless bird! Sing! Speak no more

[. . .]!"'[24] (Nietzsche 1969: 247). Yet now, when he is in this situation, Zarathustra cautions against singing *at this hour*, in 'the secret, solemn hour', the hour when no shepherd plays his pipe – because it is the hour of Pan. Now, in this hour of Pan, not even Zarathustra's soul should sing, but it, too, is to remain silent. 'Soft! The world is perfect' (*Still! Die Welt ist vollkommen*). Noontide, itself now personified, is hot and is asleep; noontide has drunk a drop of golden happiness, of golden wine. These lines reveal the profound extent to which, at its core, Nietzsche's philosophy is essentially positive: 'I should like very much to do something that would make the thought of life even a hundred times more appealing to [people]' (Nietzsche 1974: 225). At the same time, his concept of happiness is a complex one, and its aim is to discover a capacity 'for making new galaxies of joy flare up' (Nietzsche 1974: 86).

Something – what is it? – glides (*huscht*) by, just as the doves will glide (*huschte*) over the laughing lion's nose in the final chapter of the work.[25] (Nietzsche 1969: 334). What is it that glides by here? It is happiness (*Glück*) itself? (Ibid.: 288). Happiness (*Glück*) *is* this gliding, *is* 'precisely the least thing, the gentlest, lightest, the rustling of a lizard, a breath, a moment, a twinkling of the eye' *is* the moment when the world has 'become perfect'. Later on, in 'The Intoxicated Song' – not at midday, but at midnight, and drinking not wine, but dew – Zarathustra will rediscover the moment when the world becomes perfect, and recall this moment, the moment when he says, 'you please me, happiness, instant, moment!' ('*du gefällst mir, Glück! Husch! Augenblick!*') and, in wanting this moment to return, in wanting this 'one moment twice' (*Ein Mal Zwei Mal*), wanting everything (*Alles*) to return:

> What do you think, you Higher Men?
> *For all joy wants – eternity!*
>
> (Ibid.: 332)

'Thus – does a god laugh': The laughter of a god that precedes his happiness, – in the words of *The Gay Science*, 'the happiness of a god full of power and love, full of tears and laughter' (Nietzsche 1974: 268) when the evening sun begins to set, a divine feeling of a new humanity; in the words of *Beyond Good and Evil*, the 'halcyon smile' of 'the tempter god' when he demands that this new humanity become 'stronger, more evil, more profound; also more beautiful' (Nietzsche 1968: 426); the humanly-divine happiness of Dionysos. Thus – does a god laugh; thus, too, does the transmogrified Zarathustra – the *Übermensch* of the *Augenblick* – at the conclusion of 'Of the Vision and the Riddle', the laughter of 'a transformed being, surrounded with light, *laughing*!' (Nietzsche 1969: 180).[26] 'Happiness; how little attains to happiness!' says Zarathustra, alluding to an aphorism in the second volume of *Human, All Too Human*.[27] Yet Zarathustra now retracts this statement, and describes it as a blasphemy. For now, he understands indeed how truly little is required for happiness, an insight related to the insistence by Johann Wolfgang Goethe (1749–1832) on the importance of 'lesser things' or what Nietzsche called the 'nearest things'.[28] Zarathustra has learnt that '*little* makes up the quality of the *best* happiness'.

Halfway through his dialogue with his soul, Zarathustra repeats his opening question – 'What has happened to me?' – and his injunction to pay attention: 'Listen!' (*Horch!*). In this, the highpoint of his experience, it seems that time itself has stopped: or, in the imagery Nietzsche uses, time has flown away, and Zarathustra himself has fallen into the well of eternity. (In 'The Intoxicated Song', Zarathustra will ask again: 'Woe is me! Where has time fled? Did I not sink into deep wells?' (Nietzsche 1969: 328).[29] In 'At Noontide', it is that Noon that sleeps, while in 'The Intoxicated Song', it is the world that is asleep.) Again, and more urgently, Zarathustra here repeats his question: 'What is happening to me?' – and an injunction to pay attention: 'Listen!' (*Still!*). For to Zarathustra's happiness is now added to his pain – a stinging pain, a pain that stings his heart. Yet this pain has its place: It forms part of the process of fermenting and distilling the essential Zarathustrian joy – for Dionysos, the vintner god, the god with the diamond-studded vine-knife, must cut the vine.

Now this ripeness, this readiness for harvesting, has been attained by the world itself, as Zarathustra again reminds us: 'Has the world not just become perfect?' Indeed, it has become perfect, it has become 'round and ripe' (*rund und reif*), recalling the 'ripe, golden apple' in 'Of the Three Evil Things'; its ripeness (*Reife*) is its roundness (*reif*), it has become a 'golden, round ring' (*des goldenen runden Reifs*): The ancient geometrical image of perfection, the circle, which signifies the circularity of eternal recurrence. The ring of recurrence – 'the wedding ring of rings – the ring of recurrence', with which Zarathustra proposed to be wed to Eternity – (Nietzsche 1969: 244)[30] has been anticipated in Part One by the circling eagle (*in weiten Kreisen*), flying with the serpent coiled round its neck (*geringelt*) (Nietzsche 1969: 52–53)[31] as by the serpent encircling the sun on the staff which Zarathustra's disciples give him, 'a staff, upon the golden haft of which a serpent was coiled about a sun . . . a golden sun, and around it the serpent of knowledge' (Nietzsche 1969: 100–101).[32] And as the world itself (or it seems) proceeds to fly away, Zarathustra stretches himself – and realizes he is sleeping (Nietzsche 1969: 289).[33]

Zarathustra tries, in a scene not entirely lacking a humorous aspect, to bestir himself. But his soul, still rapt in ecstatic contemplation, refuses: It wants to linger in its vision of the world as a round golden ball, – like the golden ball that Zarathustra threw to his friends as heirs of his goal;[34] his soul wants to stretch, to yawn, and to sigh, as it falls even further into the deep wells of infinite time. And as an indication of the extent of his alienation from his 'thief of time' of a soul, Zarathustra starts in surprise when a beam of sunshine falls onto his face and asks: 'But who are you then, O my soul?'.

A ray of sunshine falls out of the sky, out of the heavens: a symbolic location, the religious idea of which Zarathustra repeatedly parodies.[35] But the sky also has a specific set of connotations in the rest of *Thus Spoke Zarathustra*, symbolizing openness, cheerfulness and freedom. In his discourse 'On the Blissful Islands', for instance, Zarathustra addresses his friends when it is 'autumn all around and clear sky and afternoon' (Nietzsche 1969: 109);[36] in 'Of Voluntary Bliss' Zarathustra speaks to his conscience, rejoicing with it to be 'again alone and willingly so, alone with the pure sky and the open sea; and again it is afternoon around me' (Ibid.: 181);[37] before he turns in the next chapter, 'Before Sunrise', and addresses the sky directly: 'O sky above me! O pure, deep sky! You abyss of light!' (Nietzsche 1969: 184).[38] It is in this chapter that Zarathustra gives his blessing, a blessing in the following terms

> [. . .] To stand over everything as its own sky, as its round roof, its azure bell and eternal certainty [. . .].
>
> For all things are baptized at the fount of eternity and beyond good and evil [. . .]
>
> I set this freedom and celestial cheerfulness over all things like an azure bell [. . .]. (Nietzsche 1969: 186).[39]

Using the same invocation, Zarathustra now speaks again directly to the sky, and asks the following questions:

> When will you drink this drop of dew that has fallen upon earthly things – when will you drink this strange soul
>
> – when, well of eternity! serene and terrible noontide abyss! when will you drink my soul back into yourself?
>
> (Ibid.: 289)[40]

Like other parts of *Zarathustra*, this chapter ends with a series of questions;[41] and, like those other parts, the answers to them can be found elsewhere in *Zarathustra* itself. So the answer to the question of when all these things will happen is as follows: at midnight in 'The Intoxicated Song'; when the dew falls (§10); when Zarathustra again has sunk into deep wells (§4); when 'the world itself has grown ripe' (§6); when the world again becomes 'perfect' (*vollkommen*) (§7); when – after 'youth' and 'noontide' and 'afternoon' – there come 'evening and night and midnight' (§8); when 'all joy wants the eternity of all things, wants honey, wants dregs', and when 'the will of the ring wrestles within it' (§11); when Zarathustra cries: 'O happiness! O pain! Oh break, heart!' (§11) – for 'midnight is also midday' (*Mitternacht ist auch Mittag*) (§10).

If noontide is the question, midnight is the answer, because ultimately each is the other and both are the same. How can this be? One answer is that the identity of midday and midnight recalls the initiatory moment into the Eleusinian Mysteries as described by the rhetorician and philosopher Apuleius (c. 124–170) in his novel *Metamorphoses*, referred to by Saint Augustine as *The Golden Ass*. In book 11, Apuleius writes: 'I approached near unto hell, even to the gates of Proserpine, and after that I was ravished through all the elements, I returned to my proper place' and 'about midnight I saw the sun brightly shine, I saw likewise the gods celestial and the gods infernal, before whom I presented myself and worshipped them' (Apuleius 1922: 581). If Apuleius approaches the confines of death, Zarathustra falls into *the well of time*; if Apuleius sees the sun shining at midnight, Zarathustra realizes that Midday is the same as Midnight; if Apuleius worships the gods, Zarathustra hears a god – laughing.

Moreover, the midnight sun is a motif in alchemy, symbolizing 'the spirit in man shining through the darkness of his human organisms' (Hall 2003: 145). In alchemy, the midnight sun or black sun (*soler nigus*) belongs to the first stage of the alchemical opus, and is referred to in such works as *Philosophia reformata* (1622) by the alchemical writer Johann Daniel Mylius (c. 1583–1642) (in *Psychology and Alchemy* and *Mysterium coniunctionis*, Jung specifically refers to the image as found in Mylius[42]) the anonymous *Splendor solis* (1532–1535), or *Liber de arte chemica* (1702), ascribed to the Italian Renaissance scholar Marsilio Ficino (1433–1499).

For Jung, the existence of the black sun – or the accompaniment of the sun by its own shadow – reflects 'the duality of [the alchemist's] unconscious assumptions' and 'the antinomian thinking of alchemy', inasmuch as 'while the Sol is the most precious thing, its shadow is *res vilissima* or *quid vilius alga* (more worthless than seaweed)' (Onfray 2009: 56). Finally, according to Manly P. Hall, the midnight sun referred to '"the *spiritual* sun" – as opposed to the "intellectual" and the "material" suns – in the solar system, which the mystic could see as well at midnight as at high noon' (Hall 2003: 145), thus reinforcing the validity of Jung's reading of this passage as suggesting that Nietzsche himself had experienced what the initiate at Eleusis had.

Yet neither Nietzsche nor Jung is trying to revive an ancient mystery cult (although, so far as Jung is concerned, it is no secret that Richard Noll would disagree). Yet in his account of Zarathustra's mystical experience in 'At Noontide', which stands in a complex relation to Nietzsche's own account of the experience of composing *Zarathustra*, and in Jung's analysis of it, we find a gesture towards a dimension of time which can perhaps best be described as an *Eternal Now*.

Surely this is the essence of what Jung is trying to tell us about the archaic: it precedes us, it leads us forward, it is always already there, and it can never go away. In philosophical discourse, this dimension can be traced back to Plato's definition of time in the *Timaeus* as an image of eternity (Plato 1989: 1167);[43] but Nietzsche's description in *Zarathustra* of the 'midday vision' is, like the intuitive experiences of other writers in caves, in the midst of forests and maybe on cliffs, a powerful evocation of a moment in which *the Now is revealed as eternal.*

So what? Is this not all a game with words and concepts without any implications for the way in which we choose to lead our lives? Far from it. For, as Klages puts it, 'some people are born old, and even if they *exist* for nearly ninety years, it may be that they have not *lived* for even two minutes'; whereas, by contrast, 'anyone who dies young does not, on this account, lose the moment of eternity [*Ewigkeitsaugenblick*] to which he or she was divinely blessed to be privy!' (Klages 1929: 396).

Notes

1 J. Meades (presenter), *Magnetic North*, Episode 1, broadcast BBC 4, 21 February 2008, 21:00.
2 For further discussion of Friedrich, see W. Vaughan, *Friedrich*, London and New York: Phaidon, 2004.
3 See C.G. Jung, *Archaic Man* (1931). For further discussion of this paper and the ramifications of the concept of the archaic, see Bishop (ed.) (2012).
4 Original title: Guénon, R. (1962), *Symboles fondamentaux de la Science sacrée*, Paris: Gallimard.
5 See the website presentation of the dynamic relationship between Palaeolithic cave art and depth psychology by Ilse Vickers under the suggestive title: *Cave Art: An Intuition of Eternity*. Available at www.bradshawfoundation.com/cave_art_an_intuition_of_eternity/index.php.
6 For alternative translations and further discussion, see Porphyry, *On The Cave Of The Nymphs*, trans. R. Lamberton, Barrytown, New York, Station Hill Press, 1983; Porphyre, *L'antre des nymphes dans l'Odyssée*, trans. Y. Le Lay, Paris, Verdier, 1989 (includes G. Lardreau, 'La philosophie de Porphyre et la question de l'interprétation', 7–47); and R. Lamberton, *Homer the Theologian: Neoplatonist Allegorical Reading and the Growth of the Epic Tradition*, Berkeley, Los Angeles, London, University of California Press, 1986, 108–33, 318–24.
7 Au-dessus de ma tête, souvent, je sais le ciel vide de Dieu
Mais plein d'étoiles
Et, parmi toutes, la seule, l'unique
Le magnétisme indéfectible d'une boussole pour Être. . .
(c) Editions Galilée; reproduced with permission.
8 For further discussion, see S. Wilke, '"Verrottet, verkommen, von fremder Rasse durchsetzt": The Colonial Trope as Subtext of the Nazi-"Kulturfilm" *Ewiger Wald* (1936)', *German Studies Review*, vol. 24, no. 2 (2001), 353–76; and J. Zechner, Politicized Timber: The *German Forest* and the Nature of the Nation 1800–1945, *The Brock Review*, vol. 11, no. 2 (2011), 19–32.
9 *Faust II*, ll. 6271–6274.
10 Habermas's *Theorie des kommunikativen Handelns* was originally published in 1981; the English translation was published in two volumes by two different publishers/imprints in 1984 and 1987 respectively: vol. 1, *Reason and the Rationalization of Society*, 382 and 385; *Theorie des kommunikativen Handelns*, 2 vols. (Frankfurt am Main: Suhrkamp, 1987), vol. 1, *Handlungsrationalität und gesellschaftliche Rationalisierung*, 511–12 and 516.
11 For further discussion of the links between Freiburg and Frankfurt, see F. R. Dallmayr, *Life-World, Modernity and Critique: Paths Between Heidegger and the Frankfurt School* (Cambridge: Polity Press, 1991).
12 *Thus Spoke Zarathustra*, 'Of the Priests'.
13 For further discussion of the significance of the Eleusinian Mysteries from a Jungian perspective, see C. G. Jung and C. Kerényi, *Essays on a Science of Mythology: The Myth of the Divine Child and the Mysteries of Eleusis*, R. F. C. Hull (Trans.), Princeton, New Jersey, Princeton University Press, 1969. For more recent discussion, see Christiane Sourvino-Inwood, 'Festival and Mysteries: Aspects of the Eleusinian Cult', in Michael B. Cosmopoulos, *Greek Mysteries: The Archaeology and Ritual of Ancient Greek Secret Cults*,

London and New York, Routledge, 2003, 25–49; Jennifer Larson, *Ancient Greek Cults: A Guide*, New York and London, Routledge, 2007, ch. 6, 'Mistresses of Grain and Souls: Demeter and Kore/Persephone', 69–85; and H. Bowden, *Mystery Cults in the Ancient World*, London: Thames and Hudson, 2010, ch. 1, 'The Eleusinian Mysteries', 26–48.

14 Cf. Taylor's reference to Proclus's commentary on Plato's *Republic*, where Proclus glosses Socrates's comment, 'Every seeker after wisdom knows that up to the time when philosophy takes it over his soul is a helpless prisoner, chained hand and foot in the body' (59–60; cf. *Phaedo*, 82e; in Plato, *The Collected Dialogues*, eds E. Hamilton and H. Cairns, Princeton, New Jersey, Princeton University Press, 1989, 66).

15 See Bowden, *Mystery Cults in the Ancient World*, ch. 9, 'Dionysus', 105–36; Larson, *Ancient Greek Cults*, ch. 10, 'Epiphany and Transformation: Dionysos', 126–43; and N. Robertson, 'Orphic Mysteries and Dionysiac Ritual', in Cosmopoulos (ed.), *Greek Mysteries*, 218–40.

16 *Thus Spoke Zarathustra*, 'Of Involuntary Bliss'.

17 *Thus Spoke Zarathustra*, 'At Noontide'.

18 For further discussion, see G. Schmitt, 'Von Übermensch und Übersinn: Nietzsches Metaphorik im *Roten Buch*', in *Recherches germaniques*, Hors série Nr. 8, *Arts, sciences et psychologie: Autour du 'Livre Rouge' de Carl Gustav Jung*, ed. C. Maillard, 117–34 (2011).

19 For further discussion, see K. Schlechta, *Nietzsches großer Mittag*, Frankfurt am Main: Klostermann (1954); and P. Borgeaud, *The Cult of Pan in Ancient Greece*, trans. K. Atlass and J. Redfield, Chicago and London, University of Chicago Press (1988).

20 *Thus Spoke Zarathustra*, 'At Noontide'.

21 'For convalescents should sing; let the healthy talk' ('The Convalescent' §2; in Nietzsche 1969: 236).

22 *Thus Spoke Zarathustra*, 'Of the Great Longing'.

23 See Marx, O. and Morwitz, E. (trans.). (1949). *The Works of Stefan George*, Chapel Hill, North Carolina, University of North Carolina.

24 *Thus Spoke Zarathustra*, 'The Seven Seals'.

25 *Thus Spoke Zarathustra*, 'The Sign'.

26 *Thus Spoke Zarathustra*, 'Of the Vision and the Riddle', §2; 'Never yet on earth had any man laughed as he laughed! / O my brothers, I heard a laughter that was no human laughter,' Zarathustra tells his audience aboard the ship. See Hough, *Nietzsche's Noontide Friend*, 94.

27 Entitled *Modesty of man*, the aphorism runs: 'How little pleasure most people need to make them find life good, how modest man is' (*Human, All Too Human*, vol. 2, *The Wanderer and his Shadow*, §15; in F. Nietzsche, *Human, All Too Human*, trans. R. J. Hollingdale, Cambridge, UK, Cambridge University Press, 308 (1980).

28 See Goethe's review of Ernst Stiedenroth's *A Psychology in Clarification of Phenomena from the Soul* (1824); in J. W. von Goethe, *Scientific Studies* [Goethe Edition, vol. 12], New York: Suhrkamp Publishers, 45–46 (1988).

29 *Thus Spoke Zarathustra*, 'The Intoxicated Song', §4.

30 *Thus Spoke Zarathustra*, 'The Seven Seals', §1.

31 *Thus Spoke Zarathustra*, 'Zarathustra's Prologue', §10.

32 *Thus Spoke Zarathustra*, 'Of the Bestowing Virtue', §1.

33 *Thus Spoke Zarathustra*, 'At Noontide'.

34 See 'Of Voluntary Death': 'Truly, Zarathustra had a goal, he threw his ball: now may you friends be the heirs of my goal, I throw the golden ball to you. / But best of all I like to see you, too, throwing on the golden ball, my friends!' (Nietzsche 1969: 99).

35 See, for example: 'It is not your sin, but your moderation that cries to heaven [. . .]!' ('Zarathustra's Prologue' §3, in Nietzsche 1969: 43); 'They say their marriages are made in Heaven. / Well, I do not like it, this Heaven of the superfluous!' ('Of Marriage and Children', 95); and: 'Truly, their Redeemers [i.e., the priests'] themselves did not come from freedom and the seventh heaven of freedom!' (*Of the Priests*, 116).

36 *On the Blissful Islands*. For further discussion of this chapter, see my forthcoming study, *On the Blissful Islands: With Nietzsche and Jung in the Shadow of The Superman* (forthcoming, Routledge).
37 *Thus Spoke Zarathustra*, 'Of Involuntary Bliss'.
38 *Thus Spoke Zarathustra*, 'Before Sunrise'.
39 *Thus Spoke Zarathustra*, 'Before Sunrise'. On the significance of this blessing, see L. Lampert, *Nietzsche's Teaching: An Interpretation of 'Thus Spoke Zarathustra'*, New Haven and London, Yale University Press, 176 (1986).
40 *Thus Spoke Zarathustra*, 'At Noontide.'
41 Compare with question in 'Of the Vision and the Riddle' §3 (Nietzsche 1969: 180).
42 See *Individual Dream Symbolism in Relation to Alchemy*, in CW 12 §140; and *Mysterium coniunctionis*, CW 14 §21 fn. For further discussion, see S. Marlan, *Black Sun: The Alchemy and Art of Darkness*, College Station, Texas, Texas A & M University Press, 2005.
43 I plan to discuss further the philosophical dimensions to the Eternal Now in 'At Noontide' in a work provisionally entitled *Now, Now: The Great Noontide.*

References

Apuleius. (1922). *The Golden Ass*, trans. W. Adlington, rev. S. Gaselee, London, Heinemann; New York, Putnam

Bishop, P. (ed.) (2012). *The Archaic: The Past in the Present*, London and New York, Routledge

Emerson, R. W. (1983). Chapter 1, 'Nature'. In J. Porte (ed.), *R.W. Emerson, Essays and Lectures*, New York, Library of America

Evers, T. (1987). *Mythos und Emanzipation: Eine kritische Annäherung an C.G. Jung*, Hamburg, Junius

Freud, S. (1961 [1930]). Civilization and its discontents. *The Future of an Illusion, Civilization and its Discontents, and Other Works*, Standard Edition of the Collected Works, Vol. 21, ed. J. Strachey and A. Freud, London, Hogarth Press, Institute of Psycho-Analysis

George, S. (1949). *The Works of Stefan George: Rendered into English*, trans. O. Marx and E. Morwitz, Chapel Hill, North Carolina, University of North Carolina

Giegerich, W. (2010). '"Liber Novus," that is, the new Bible: a first analysis of C. G. Jung's Red Book'. *Spring: A Journal of Archetype and Culture* 83: 361–411

Goethe, J. W. von (2001). Faust II. In *Faust: A Tragedy*, ed. C. Hamlin, trans. W. Arndt, 2nd edn., New York and London, Norton

Guénon, R. (1995). *Fundamental Symbols: The Universal Language of Sacred Science*, ed. M. Vālsan, trans. A. Moore (revised M. Lings), Cambridge, Quinta Essentia

Habermas, J. (1984/1987). The theory of communicative action, trans. T. McCarthy, London, Heinemann; Cambridge, Polity Press, vol. 1, *Reason and the Rationalization of Society* (382, 385); *Theorie des kommunikativen Handelns*; 2 vols. (Frankfurt am Main, Suhrkamp, 1987), vol. 1, *Handlungsrationalität und gesellschaftliche Rationalisierung*, 511–12, 516

Hadot, P. (2001), *La Philosophie Comme Manière de Vivre: Entretiens avec Jeanne Carlier et Arnold I. Davidson*, Paris, Albin Michel

Hadot, P. (2011). *The Present Alone Is Our Happiness: Conversations with Jeannie Carlier and Arnold I. Davidson*, trans. M. Djaballah and M. Chase, Stanford, California, Stanford University Press

Hall, M. P. (2003). *The Secret Teachings of All Ages*, New York, Tarcher/Penguin

Homer. (1993). *The Odyssey*, ed. and trans. A. Cook, 2nd edn., New York and London, Norton

Horkheimer, M. and Adorno, T. W. (1996). *Dialectic of Enlightenment: Philosophical Fragments*, trans. J. Cumming, New York, Continuum

Hough, S. (1997). *Nietzsche's Noontide Friend: The Self as Metaphoric Double*, University Park, Pennsylvania, Pennsylvania State University Press

Jung, C. G. (1923). *Psychological Types*, in CW Vol. 6, H. Read, M. Fordam and G. Adler (eds.), trans. J. V. Heurck, London, Routledge & Kegan Paul

——. (1927/1931). *Analytical Psychology as Weltanschauung*, in CW Vol. 8

——. (1931). *Archaic Man*, in CW Vol. 10

——. (1940/1950). *Concerning Rebirth*, in CW Vol. 9i

——. (1988). *Nietzsche's 'Zarathustra': Notes of the Seminar given in 1934–1939*, J. L. Jarrett (ed.), London, Routledge

——. (2009). *The Red Book: Liber Novus*, ed. S. Shamdasani, trans. M. Kyburz, J. Peck, and S. Shamdasani, New York and London, Norton

Klages, L. (1929). *Sämtliche Werke*, 1964–1992, Vol. 3, *Philosophie III*, eds E. Frauchiger, G. Funke, K. J. Groffmann, R. Heiss, and H. E. Schröder, Bonn, Bouvier

Klages, L. (1929). *The Science of Character*, trans. W. H. Johnston, London, Allen and Unwin

Knight, J. (1936). *Cumean Gates, a Reference of the Sixth Aeneid to Initiation Patterns*, Oxford, Blackwell

Nietzsche, F. (1917). *Thus Spoke Zarathustra*, trans. T. Common, New York, Boni and Liveright

——. (1968). *Basic Writings of Nietzsche*, ed. and trans. W. Kaufmann, New York, The Modern Library

——. (1969). *Thus Spoke Zarathustra*, trans. R. J. Hollingdale, Harmondsworth, Penguin

——. (1974). *The Gay Science*, trans. W. Kaufmann, New York, Vintage

——. (1888/1968). Twilight of the idols. In *The Anti-Christ*, trans. R. J. Hollingdale, Harmondsworth, Penguin

Onfray, M. (2009). *Le Recours aux Forêts: La Tentation de Démocrite*, Paris, Galilée

Otto, R. (1958). *The Idea of the Holy: An Inquiry into the Non-Rational Factor in the Idea of the Divine and its Relation to the Rational*, trans. J. W. Harvey, London, Oxford, New York, Oxford University Press

Ovid. (1987). *Metamorphoses*, trans. A. D. Melville, Oxford and New York, Oxford University Press

Plato. (1989). *The Collected Dialogues*, eds E. Hamilton and H. Cairns, Princeton, New Jersey, Princeton University Press

Porphyry. (1999). The cave of the nymphs. In J. Gregory, *The Neoplatonists: A Reader*, 139–146, London, Routledge

Seneca. (1932). *Epistulae ad Luciulium*; Seneca, *Letters to Lucilius*, trans. E. P. Barker, Oxford, Clarendon Press

Taylor, T. (1995). A dissertation on the eleusinian and bacchic mysteries [1790/1816]. In *Oracles and Mysteries* [Thomas Taylor Series, vol. 7], On the Eleusinian Mysteries, 59–127, Frome, Prometheus Trust

Thoreau, H. D. (1908). *Walden; or, a Life in the Woods*, London and Toronto, Dent; New York, Dutton

Thoreau, H. D. (2001), *Collected Essays and Poems*, New York, Library of America

Vaughan, W. (2004). *Friedrich*, London and New York, Phaidon

Wegmann, P. (1993). *Caspar David Friedrich to Ferdinand Hodler: A Romantic Tradition: Nineteenth-Century Paintings and Drawings from the Oskar Reinhart Foundation, Winterthur*, Frankfurt am Main und Leipzig, Insel

8

MOTION, MEASUREMENT, AND MECHANIZATION

The sacred-secular in temporal relics, objects, and possessions

J. A. Swan

A Dream: Gnomon.

> A man stands in an empty field. His figure is circled by a ring of stones. The shadows of the stones have grown and diminished with the light energy from the nearest star mapped onto the celestial charts: behind him, then above him, then beginning its descent, as it had arisen. He has lived half of the years which comprise an average lifespan expectancy. He is at Mid-Life. He begins to walk again; the clock upon his wrist is wound by this motion. The metal plate sealing the mechanical gears presses against his skin. His name and the date of his birth are engraved beneath the name and date of the previous generation, and the one before.

Architectural structures, artefacts and personal possessions are material entities indentured to human time. They share a curious simultaneity: as inanimate functions of form dedicated to their own purpose, and, as temporal vessels interacting with meaning and the capacity to hold onto histories and cultures, past and present. Their *libido* is not human, but of humankind. Perceptions of life energy in the inanimate state are invoked by the quality of human interaction which takes place between man and his objects: the evolutionary requirements of practical conditions, the desire for contact with beauty, the placating of the ego and its consumptive wants, the seeking of historical truths, and the need for factual revelations about the history of human life (from whence they came, and who their makers were).

The inanimate state of being fuses a particular object's origin (or, inception) into a specific historical time, while permitting the object to exist as symbolic information remaining *within* linear time. Inanimateness anchors an object's historical information in such a way that it is only accessible – even discernible – through

FIGURE 8.1 *Natural Time*, 2016 (J. A. Swan).

remaining within the process of human time. In this way, objects become a temporal expression of matter, illuminated through the experience of contact with human beings, their myths, their technologies.

From raw matter origins of stone and wood, glass and metal, fibres and pigments, an object's lifetime begins with the transformation process: Objects are recomposed by creativity and culture into a whole new presence within the material world; objects that began their life journey not as a cellular being, but as an idea or image are formed with materials and the creative spirit of the human mind. As matter, objects engage in the interactions and experiences of the human lifespan, they are neither human nor animate, and yet, they depend upon the natural presence of humankind to exist in form. As creative thoughts transformed into a physical presence, these objects become specific – even unique – place markers within the culture and conditions of a particular human time.

More so than any dying mythology or transient physical landscape in this post-Modern age it is the material objects that we inherit, acquire, collect, surround ourselves among, and live our lives within that hold our collective historical imprint and identification within time. They are a physical conduit to our own unique and personal timelines, facilitating the capacity for *memory recollection through objects* about a particular time passing in our lives as we change, and change again, moving forward.

Life in materials

In linear progression, *inanimate time takes time to observe*. Hodder (2012: 4–5) clarifies that material objects are not – by way of physical composition – in and of human time: They are parallel, invisibly chemically transforming at the atomic level and, 'endure over different temporalities.'[1] The capacity for temporal endurance – embracing the stretching of time – is what distinguishes material objects from their makers. In this parallel timeline, inanimate objects hold their own fluid sense of lifespan. Images become symbols, move between connotations of sacred and secular life, change their physical character of form, and adjust to their interpreted meanings.

Such is the fluidity of time as we, too, move through our animated lives: Our biology and cultural conditions evolve, our hours and seasons journey from darkness into lightness, over and again, and we grow into our changes, transforming willingly or by force of natural circumstances. *We become whole in time*. Within the temporal expression of this process, a pattern emerges: a circle, a cycle.

The image of the circle is among the most ancient of symbols to appear by way of human beings, throughout millennia. It is, at once: the celestial bodies, the Gods, the womb, a reflection of wholeness – *it is what we are becoming*. As an image, the circle is an organic shape, a shape that is *of nature*; Freed from angularity and the conflicting of points wherein the direction of its line flows fully round, and again, until the line is broken or opened, or the cycle of its life ends. It is the physical shape that human beings most associate with Time and time-keeping whether observing the motion of the celestial bodies, securing the architecture of the standing stones, or adjusting the mechanization of clockworks, in towers or houses and upon our wrists, born from a most recent Modern age.

This chapter explores the developmental process of a particular set of material objects: those dedicated to the observation, recording, and measurement of Time. With an emphasis upon transformation through the materials of composition, the practicalities of function and the aesthetics of form, it is a story about both cyclical and linear progression, and adaptation of physical objects to accommodate emerging culture and changing technology. It is also a story of the symbolic: Observing the patterns and rhythm of our *lives-in-time* and how we have translated our internal perceptions of temporality into the material world.

A becoming cycle

> While I was caught up in this dream of a static, age old existence, I suddenly thought of my pocket watch, the symbol of the European's accelerated tempo . . . His watch tells him that since the 'Middle Ages' time and its synonym, progress, have crept up on him and irrevocably taken something from him . . . He compensates for the loss of gravity and the corresponding sentiment d'incompletude by the illusion of his triumphs . . . such as steamships, railways, aeroplanes, and rockets, that rob him of his duration and transport him into another reality of speeds and explosive accelerations.
>
> (Jung 1995: 268)

Jung frequently made the observation that human beings are part of a living collective awareness which cycles unconsciously, actively engaging and re-engaging the psychic imprint of historical millennia in an *atemporal plane*.[2] This

FIGURE 8.2 Earliest mechanization for a 'Self Winding' wristwatch, 1928.

(London: Museum of the Worshipful Company of Clockmakers, est. 1631)

interaction is expressed through the continual emergence of symbolic patterns, our psychic vistas through the cosmic window to an understanding of human life: *Who We Are (Now)*, *Who We Have Been*, and developing thoughts about *Who We Are Becoming* are temporally challenging questions that illuminate the process of *changing within* our lifetime as individuals, and collectively, across millennia as human beings. Central to this *becoming* process is the organic connection – the essence of *life-contacting* – within and through the Unconscious, and the current, or motion, with which human growth and psychic development occurs. The individuality potential inherent in each unique lifespan as it connects at the deepest, archaic level of cosmic time is our *natural* condition within *becoming* human. Jung's heliocentric analogy of Life-Time stages[3] unifies the organic motion of human development and the cycle of the natural environment within cosmic time: 'Time is . . . defined by the rising and setting sun, by the death and renewal of the libido, the dawning and extinction of consciousness.'[4, 5]

Central to the heliocentric model – in both sacred systems of belief and secular applications of science – is the tangible, observable quality of cosmic time: the celestial bodies in motion provide a physical sense of predictability that remains a constant against mutable human conditions and environments. We ascend and descend through the biology of a linear lifespan as we transform interiorly within an atemporal plain.

Along the developmental path – which has taken more than forty millennia (Donald 1991; Griffin and Tyrrell 2011; Mithen 1996, 2004; Renfrew 1998) – we have acquired the consciousness and means of creative reflection about our relationship as individuals to cosmic time. We have come to understand – as evidenced from the earliest incarnations of the mythological 'Fall'[6] into human time – that we are, at once, in and of a specific *lifetime* with its individual and collective connotations, and that we are powerless against the motion of cosmic time. The passivity of our observational state is not without change, however: We have learned intuitively[7], and later, for practical reasons to be time-keepers with the knowledge that *keeping in time* with the world around us provides a structure or scaffold upon which cultural (and, individual) development progresses. As such, our time-keeping skills have transformed through the creative spirit that underpins the physical emergence of archaeoastrological[8] sites, temporal instruments and time objects.

Much of what we know about our *past times* and our linear development through Time comes from architecture, artefacts and relics: sites and objects that are no longer required, left behind in haste, or were perhaps passed along generations until the passing into Time became obsolete. What remained of these 'things'[9] became static in the state of disuse, and eventually, their cognitive life – the interaction with human kind – stopped at an *exact* temporal and physical location, waiting for rediscovery and re-contextualization within a new and different lifetime. *The dormant object libido is re-awakened with human contact.* Our historical relationship with physical matter – our *material lives* in the natural world – thus provides an exhibition gallery for understanding how the Past fits within our Present human time.

Time in mind

Scholarship has established the connection between the evolution of cognitive function and the creative process of imagination required for the transformation of matter-into-form to convey abstract observations (Malafouris and Renfrew 2010; Mithen 1996, 2001; Orians 2001): The imaginative thought that the same matter (here, stone) could be used as carving tool, a hunting spear, and was equally capable of, and used for, a temporal record-keeping device detailing the lunar cycle. Frank illustrates that:

> We can call this poorly understood but essential connection between the physical world and cultural invention *enigmatic entanglement*. Through this enigmatic entanglement, a remarkable dialogue between mind and matter was begun – forever linking cosmic and human time together . . . In this way, time was a creation of culture, just as culture was a creation of the embodied mind. The enigmatic entanglement between experienced physicality and the symbolic, cultured imagination created a time that existed in both daily life and myth.
>
> (Frank 2011: 1–11)

The cognitive understanding of 'context beyond the moment' (Griffin and Tyrrell 2011: 24) developed through external observation of the cycles of cosmic time and the elements of the natural world, and was perhaps solidified from within through the process of dreaming, specifically, daydreaming: 'What we believed occurred about 40,000 years ago is that humans learned to access the REM dream state while awake.' (Ibid.: 20). *They were, as we are now, dreaming the myth onwards.* Griffin and Tyrrell remark upon our ancestors' developing a temporal consciousness:

> As well as concentrating on the present moment, they could focus on the past or the future and what was present or not present. More importantly, they could choose to draw on their past experiences to see how the present is embedded in a deeper context, and because of that, how any changes they might make could affect the future.
>
> (Ibid.: 70)

The acquisition of a temporal consciousness created the need – or even, *a psychic drive* – to physically observe, and later, record the presence of time. The earliest time recording devices – portable calendars [10]– would have been *carried around* as a life object twenty millennia before human beings made even the most rudimentary of Neolithic settlements.[11] Our similar need *to observe, to display, to carry* or, *to hold* (unto) time remains apparent. There is a culminating sense of 'unification'[12] within our contact with the psychoid nature of Time, the desire or drive for contact with the *materialization of temporality* across our history of engagement with time objects: The Palaeolithic bone etched with a cosmic cycle, the astronomical sky ceiling, the sundials and clepsydras of antiquity, the Medieval clock tower, the mechanical pendulum table clock, the gilded Victorian pocket watch, the 20th Century wristwatch, the futuristic carbon fiber 'all-in-one' time *and-everything-else-about-our-life* keeper. Perhaps the differentiation among these objects is found within the psyche's connection to what Time *is and does*: Its sacred and secular connotations; its individualistic and collectivistic experiences; its material and spiritual expressions. And, how these characteristics emerge through the materialization of objects we have brought forth creatively through developing culture in history.

Observing records

As human beings, we connect to our external life through observing and recording information about the worlds around us. We organize these perceptions to understand how it is we are *becoming* within our own time. Organizing our recorded

FIGURE 8.3 Celestial Planisphere, Babylonian (Nineveh) 650 BC, baked clay.
(London: The British Museum)

observations of cosmic time is inexorably linked to the emergence of formal writing,[13] which brought forth through enigmatic entanglement the creative invention of the calendar and its variations across the ancient developing world. In discussing temporal objects, our knowledge about past times – beyond the analysis of objects themselves, and especially, prior to the *Radiocarbon Revolution*[14] – relates to analysing recorded details about the convergence of cosmic and human time in the form of calendars. The thought persists that, the calendar itself is perhaps the most obvious temporal object as it incorporates not only the features of day-to-day living, but also, the collectivity of recorded conscious lifetime as experiences stretch from days-into-months-into-years, over and again as the cycle repeats.

> Since most calendars are based on astronomical cycles . . . they provide a link between mankind and the cosmos and can reach sacred status. Without a calendar . . . it is very difficult to memorize the past in a structured way or to plan life events, be they farming and hunting or religious and sacred duties. Inaccuracies in a calendar, which bring it out of step with its astronomical basis, can seriously upset the life of a society, while reforming a calendar which people have become used to and which is part of their everyday lives is an extraordinary event.
>
> (Lippincott 1999: 48)

The origin of the calendar as a temporal object – and the inception of the calendar-keeping process – is not without significant complexity and fluidity in its evolution, from the earliest incarnations in the Ancient Near East to the familiar Gregorian system that unifies contemporary secular life, internationally (commercial business and industry, and travel, in particular, rely on the synchronization of time across cultures). The history of this developmental process is, however, rendered unconscious in our contemporary interactions with the calendar as a tool for the organization of our personal life within linear time. Like mechanical time, it has become part of the non-negotiable individual lifetime-keeping apparatus, a physical extension reflecting our whole life as we live as Modern human beings. The calendar – like all temporal objects – functions as a life recorder and unifier. We are unified by our human connection to natural time, and the essence of the calendar provides a visual and mathematical foundation, a metric clarification through image of *extended time* for a physically unobservable linear process.

All complex calendar systems have at their centre the cosmic cycle, and as such, the archaic imprint of time. This recorded observation of cosmic time has been, historically during the pre-Christian era, a bi-linear model incorporating both the lunar and solar cycles, a *lunisolar* calendar. An exception to this is the solar calendar of Ancient Egypt,[15] from which the (Christian) Julian, and later, Gregorian calendars arise. While the Islamic Hijra calendar, instituted most recently

FIGURE 8.4 Portable sundial.

(London: National Maritime Museum, Greenwich)

in 622 AD, remains the only lunar calendar system widely instituted, at present. In both instances, the historic transformation to a monotheistic model of faith is reflected in the singularity of the cosmic cycle imprint.

The cosmic cycle, faith-based celebrations underpinning our calendar systems present a curious cultural paradigm in the contemporary world: An intra-cycling of individual calendars is active, and activated, dependent upon one's personal belief system, geographic location and the predominant religious faith (if any) of that region. In this paradigm, variants of sacred and secular life are in direct contact (and often opposition) with one another through the intra-cycling of two (or more) temporal record systems developed for organizing the practice of beliefs. The experience is one of simultaneity within the collectivity of environment.

As such, calendars are an exceptional example of temporal objects in this discussion of the mechanics of time. They are completely reliant upon the intrapersonal contact with human beings. Their *libido mechanism*, to provide an account and description of natural cycles and human life, requires direct human input, ongoing. In this way, calendars bridge the connection between human beings and the temporal independence of natural and mechanistic time-keeping objects.

Natural motion of time

> [A]ll the clocks used by man, at least until the invention of mechanical time pieces, were in their way, linked to our bodily location. Time was measured against the visible motion of the stars and the 'rising' and 'setting' of the sun, that is movements that exist only in relation to our point of view.
>
> (Umberto Eco, in Lippincott 1999: 14)

It is perhaps not a coincidence that the two most prevalent types of time-measuring devices in antiquity reflect the cosmic symbolism of the lunisolar calendar: The water clocks (or, clepsydras) and solar clocks (or, shadow clocks and sundials). For while the ongoing human process of observing and recording *the cosmic times* was the remit of the ancient calendar, the everyday cycle of life underpinning calendar-keeping required further delineation, and acquisition of (relative) accuracy. This distinction of *days in time* defined a secondary, secularized organizing of the life activities brought about by and necessitated though developing culture and increased urbanization. In the ancient world, natural time-keeping emerged in devices to clarify and support the developing calendar system, creating a temporal cycle-within-a-cycle, a macro-micro continuum, with human beings at the centre of the sphere. (The imprint and motion of this cycle-within-a-cycle remains today, albeit with new and different forms of time-keeping devices, objects that reflect and contain an all-in-one approach to timing the calendars of our individual and collective lives.)

The use of nature's mutable essence to measure a process as complex as the human experience of time has limitations if one is focused upon obtaining only precision with each passing temporal increment. This is not to say that natural clocks were incapable of improvement, or refinement. They are bound into the evolutionary process through the enigmatic entanglement of matter-within-life, and the Ancient Greek and Chinese cultures provided a place and space for the significant technological progress in the mechanics of solar and water temporal objects, respectively. It is the nature of the psyche, however, *to continue*, and this progression requires re-adjustment from time-to-time through change in view point and accommodation for new connections to external life, ongoing.

The visual symbolism apparent in solar clocks, in particular, remains unsurpassed in the developing model of our heliocentric lifetime: The mechanic styling of the sundial's aesthetic form requires an upright, solar shadow-casting

FIGURE 8.5 'The Great Sloane Astrolabe', ca. 1295 AD.

(London: The British Museum)

instrument, or gnomon – literally, 'one who comes to know'.[16] Time is understood through the gnomon whose projection is relative to its position on the earth, and within the cycle of the sun.

The conscious mind often knows little or nothing about its own transformation, and does not want to know anything. The more autocratic it is and the more convinced of the eternal validity of its truths, the more it identifies with them. Thus the kingship of Sol, which is a natural phenomenon, passes to the human king who personifies the prevailing dominant idea and must therefore share its fate. In the phenomenal world the Heraclitan law of everlasting change prevails; and it seems that all the true things must change and only that which changes remains true (Jung CW 14: 503).

Ages of time

Historians, as cultural time keepers, have the ongoing process (or, precisely, a process-within-a-process) of debating about the accuracy of *time-within-times*. The historical time measurement of European culture, for example, continues to focus upon devising a cohesive system to describe the accurate linear path of cultural change following perhaps the second most significant *Fall*, that of the Western Roman Empire in, 476 AD. What is observable from a temporal perspective is this: That a length of one millennia – characterized by a diminishing sense of collective unification, a *disunified* time – followed this pivotal event; and, that the observation of technological progress – with respect to temporal instruments – incorporated a clear distinction between the physical emergence patterns in the Near East, and the European Continent.

The temporal symbolism however, remains unchanged, a cyclical reflection of archaic underpinnings. The astronomical calendars of antiquity unified the earliest recorded observations of the various celestial cycles into a lunisolar model. The physical and metaphorical spiritual division of Europe and the Near East during the Middle Ages so too appeared within the transformation of the calendar cycle. For while the Christian Europeans focused their temporal emphasis upon the path of Sol, the Islamic calendar aligned with Luna. The creative capacity to embrace the fluidity of the symbolic at this point in human development is significant, as the process of organizing temporal life had, by the historic birth of Christ, the collective cultural imprint of nearly four millennia of progress. This *life-contacting* with the unconscious imprint of symbolic time produced patterns of emergence in physical objects that transcended the linear development of temporal technology.

The technological advancements through linear time were still evident, however, particularly in Arabian timekeeping devices. Cultural migration and assimilation of the Ancient Greek knowledge of practical and theoretical time was a crucial transference of collective ideas. Without advancement of Arabic culture during the Middle Ages, the foundation of Ancient astronomical thought – and by extension the 17th century's developing precision in nautical time-keeping devices – might well have dissipated within time. The collective imprint of developing *time-and-place* is also a crucial element in the emergence of temporal objects in Arabian time keeping: The astrolabe is 'a model of the Ptolemaic universe in which the spheres have been flattened' (Lippincott 1999: 104). And while the mathematical foundations of this instrument reflect the Ancient Greek spirit of philosophical invention, the refinement of its time-keeping properties is in harmony with the Muslim scholars' emphasis upon precision in astronomical science. The physical form of the astrolabe experiences a geographical synchronicity: The flattened pattern of the sky references the star clocks and astronomical bas relief ceilings of the Ancient Egyptians, as well as the circular composition found in the clay tablet celestial planispheres of the Babylonians, the original mathematicians and astronomers of the ancient Near East.

Etymologically, the term, 'astrolabe' resonates with the perception of a developing temporal viewpoint: the translation from both Greek and Arabic is 'star taker' or 'taker of the stars.'[17] The thought implies a reversal in the role of human beings' historic observance of cosmic time. By the Middle Ages, the role of human positioning in the perception of time had experienced widespread collective change. Emphasis had shifted upon the human ability to harness the elements of the universe, and the possibilities inherent in this paradigm. This shift in viewpoint is significant to the 20th century development of atomic time, and it is not without the thought that the gestation of such advancement is connected culturally and geographically through the earliest astronomers of the ancient *fertile crescent.*

Structured time

> Symbolic language, which is regularly employed in myths, appears to be the most suitable medium of expressing an otherwise incomprehensible experience. Perhaps this is the reason why so many poets have dealt with the manner in which time is felt and perceived – and most likely it is why I find myself frequently quoting such poems, as they can be the most suitable medium through which these experiences can be voiced.
>
> (Yiassemides 2014: 101)

This discussion on time objects would be incomplete without a reference to a second example of how we communicate our observation of time in written language. *Books* are forgiving in their ability to encapsulate a particular time through the descriptions inherent in the poetry of words. The nature of observant language may date a work within a particular period of time; whereas, and moreover, a writer whose ability to harness a particular descriptive line may find their works resonate in a timeless fashion. Works of fiction, in particular, invite us to time-travel, and to experience a lifetime within worlds that would not necessarily be accessible, or even possible, without an invitation through the literary portal.

Umberto Eco (1932–2016) is such a writer, and his engagement with the temporal life in literature is evident in his novels, *The Name of the Rose* (1980), and *Foucault's Pendulum* (1988), among others. The former is set within an Italian monastery during the late Middle Ages,[18] and as such, it provides an introduction to the role of the Christian faith in timekeeping and temporal objects during a period of cultural and technological transition between the Ancient world and the Renaissance of Continental Europe. In his introductory *Note*, Eco informs the reader:

> Adso's manuscript is divided into seven days, and each day into [eight time] periods corresponding with the liturgical hours . . . Adso's reference to the canonical hours caused me some puzzlement because their meaning varied according to the place and season; moreover, it is entirely probable that in the fourteenth century the instructions given by Saint Benedict in The Rule[19] were not observed with absolute precision . . . The calculation [of Canonical hours] is based on the fact that in northern Italy at the end of November, the sun rises around 7:30 A.M. and sets around 4:40 P.M.
>
> (Eco 2001: 7)

Eco's *Note* evocates the thought that the temporal life of the monastery was based around the qualities of *collective conformity* and *repetition*, underscored by the 'probability' that adherence to the 'precision' of time was unlikely. Frank (2011: 71) remarks that this 'regularity and repetitiveness also demanded a submission to a temporal order that was imposed by the collective'. It would be true, too, to say that a 'temporal ordering' was *naturally* in time through the connotation of the heliocentric rhythm of division within days, and the accompanying Christian solar calendar. The emphasis of the European Christian church upon the rigid (re)structuring of time – and by extension, life-work within that structure – differs greatly from the observation of Arabic Moslem emphasis upon the innovation and exploration of time through mathematics and astronomy, during the Middle ages.

Central to Eco's plot is the cognitive life of monastic architecture, in particular, the library-labyrinth complex housed within a fortified tower. In his analysis of the canonical journey of the Christian *Shepherd of Hermas* (140 A.D.), Jung reflects upon the role of the psychoid in the symbol of the Tower and the Christian church.

> We began the vision of Hermas, in which he saw a tower being built. The old woman, who at first declared herself to be the Church, now explains that the tower is the symbol of the Church . . . for Hermas, it is only the Tower that matters . . . The detachment of the libido from the real object, its concentration on the symbol, and canalization into a symbolic function, is complete. The idea of a universal and undivided Church, expressed in the symbol of the seamless and unpregnable tower has become an

> unshakable reality in the mind of Hermas . . . Ecclesiastical psychology employs these same [archetypal] images in order to guide the libido towards a figurative object . . . using it to employ ideas that were of supreme importance for the minds of that time, namely, the consolidation and organization of the newly won Christian attitude and view of the world.
> (Jung CW 6: 402)

The *monastic model* is one of structure and containment.[20] The books are life energy, and the library facilitates the libido's *connection to continuation* through the cultural exchanges of writing and the experience of the written word characteristic of Medieval monasteries, while the tower's fortification renders the libido impervious to outside corruption during a period of uncertainty, characterized by the quickening process of collective historic change prior to the Renaissance.

Fluid changes

The concept of *nature-in-time* has a pattern of appearing in different practical circumstances, aesthetic forms, and in different geographic locations, across the history of temporal objects. Here would be an appropriate place to return to the concept of enigmatic entanglement with natural matter, for without this organic underpinning – quite paradoxically – the widespread development of mechanized time-keeping would not be possible. Much has been said already about the *solarization* of time present in the heliocentric model. But it is the physical introduction of *water* to the concept of measuring cosmic (and later, human) time that has facilitated the continuation of technological progress. It is perhaps not a coincidence that the natural element most frequently associated (metaphorically) with *continuation* of internal psychic motion – the temporal flow of the psyche, or, the *continuous flow* of the unconscious, emerges at this point in the external development of the physical process of time-keeping through enigmatic entanglement with the material world. It is a *continuation*; the *transformation of material* into a functional form, facilitated through an ever-increasingly *experienced* conscious (and, unconscious) human life wherein the capacity for change continues to arise.

> The waters symbolize the universal sum of virtualities . . . the reservoir of all possibilities of existence; they precede every form, and *support* every creation . . . immersion into water signifies regression to the preformal, re-incorporation into the undifferentiated model of pre-existence. *E*mersion repeats the cosmogonic act of formal manifestation; *im*mersion is equivalent to a dissolution of forms. This is why the symbolism of the water implies both a death and rebirth. Contact with water always brings a regeneration – on the one hand because dissolution is followed by a new birth, on the other because immersion fertilizes and multiples the potential of life.
> (Eliade 1987: 130)

Water clocks find their linear origin in the Ancient Near East,[21] and components of their linear development may be traced subsequently in the time-keeping devices of the Ancient Greeks and Romans[22] through evidence of the earliest water clocks,[23] clepsydras,[24] and anaphora[25] clocks.

A significant example of spontaneous emergence of horologium[26] is found within the use of water clocks in China, contemporaneous to their emergence in the Ancient Near East; and, subsequently, evinced in the independent Chinese development of hydraulic-powered (here, water wheel) astraria[27] towers, parallel to linear advancement of Arabic[28] and European devices already considered, during the Middle Ages.[29] Such a pattern has emerged before, considering the multi-cultural development of calendar systems, reflecting the need to record the observation of cosmic time. Time re-appears in different physical forms and across cultures, the libido of temporal objects stimulated and refined through human contact, development and progress.

In this survey of temporal objects, it has, thus far, been unusual to encounter an individual design – a singular object – that bears mentioning in such detail as the astronomical clock tower at the (Medieval) Palace of Kaifeng, designed (and built c. 1088–1094 A.D.) by Chinese scientific scholar, Su Sung (1020–1101 A.D.). This temporal object is unique in this survey, as it holds the symbolism of the present, or even, *the centre* – through its ability to visually connect the technological past, while simultaneously anticipating the future of time-keeping devices. It remains perhaps the most recognized – if not the most reproduced – example of Chinese temporal design. As a psychoid emergence, it resonates with the physical representation of the sacred-into-secular Medieval marriage between early mechanical technology and ancient systems of temporal measurement, while underscoring the presence of human observation within cosmic time. The technological progress of astronomical and mechanical time-keeping is layered within its practical and aesthetic design:

> Over thirty feet [twelve metres] high and surmounted by a bronze [water] power driven armillary sphere[30] which would have been used for astronomical observation. Below this was a rotating celestial globe – also powered to work automatically – with which comparison could be made from the observations carried out above it [on the armillary deck]. Beneath this again was a five-storeyed pagoda-like structure, each stage of this carrying a procession of automata, ringing bells and gongs and otherwise indicating the hours and special times of the days and night.
>
> (Jagger 1977: 61)

In this paradigm, the court astronomers are literally *observing from within* the mechanized reflection of cosmic time. They are positioned upon an observational platform of temporal recording instruments, simultaneously contained by the structure of a clock tower reaching into the sky. The sound of time surrounds the experience, through the movement of the open, mechanical gears, and the chiming of the bells.

In a duality of natural and mechanical timing, the recording of celestial motion in Su Sung's clock is *driven* through the advent and addition of material mechanization – an early form of escapement[31] and the first example of a transmission chain, Su Sung's *Celestial Ladder*. This drive into motion is initiated, however, by a *natural force,* that of flowing water. Su Sung reflects:[32]

> There have been many systems and designs for astronomical instruments during past dynasties all differing from one another in minor respects. But the principal use of water power for the driving mechanism has always been the same. The heavens move without ceasing but so does water flow [and fall]. Thus if the water is made to pour with perfect evenness, then the comparison of the rotary movements [of the heavens and the machine] will show no discrepancy or contradiction; for the unresting follows the unceasing.
>
> (Su Sung: 1094; trans. Joseph Needham, in Landes 1983: 8)

Unlike natural time-keepers that silently reflect the cosmic path of celestial bodies, all mechanical forms of time-keeping devices, 'utilise a constant force to drive a mechanism that can measure constant and equal intervals' (Lippincott 1999: 138). The deliberate human introduction of external matter *into time* as a 'constant driving force', is thus perhaps the most pivotal aspect of the developmental changes in the human perspective of temporality since our temporal consciousness emerged in the Palaeolithic period. From the perspective of the developing psyche, this introduction of matter into the *motion-mechanization* of Time itself shifted the quality of our connections and interactions with the archaic temporal system irrevocably. Increasingly, cultural emphasis is moved from the shared, organic observance and recording of cosmic time, to focus upon the maintenance of a new, individual and *artificial life rhythm*: one of linear 'constancy' and standardized precision found incrementally in perfected, 'equal intervals' – whether hours, minutes, or seconds – of human-mechanized time.

The deliberate introduction of man-made materials into the physical motion of time-keeping – the birth of mechanical time – initiated change in our perceptions of time relative to the cosmic cycle. Its natural rhythm no longer held as an essential component to understanding the structure of organized-lifetime intervals. Life activities brought forth through adherence to patterns (whether sacred, or secular) within heliocentric or lunar models now function secondarily. This is most significant, for while the use of calendars continues to reflect the symbolic imprint of celestial presence across both linear and cyclical time, the mechanization process has presented us with a choice, in everyday lifetime: whether or not *to follow the stars*.

FIGURE 8.6 *Pilgrim Postcard*, Cathedral of St. Andrew at Wells, Somerset, UK. c. mid-20th Century, Collection of the Author (Photo: J. A. Swan 2016)

The clock house[33]

The transmission, or translating, of temporal objects – and their technology – across geographic locations, and through various languages and cultures, incorporates both linear and non-linear motion. Of the former, a linear projection is visually preferential when considering the ways (and when) to organize and to discuss the history of Human Time. The latter, however, is less apparent visually: communicating *the emergence reflection of the organic pattern of Time itself.* Thus, research evidence to support a psychoid connection at the collective level – transcending culture, or even, linear time – while plausible, does not appear as frequently in research about time objects and the emergence or advancement of temporally driven technology. This is curious, considering that the image of Time is a circle, and that scholarship by its definition is bound to organizing the facts of history – irrespective of the visual patterns in emergence – to illuminate our collective Past Times. It comes without surprise, then, that the way in which Europe came to have an astronomical clock tower in nearly every major city – without clear evidence of a series of trial-and-error architectural stages – by the end of the Middle Ages,[34] remains debatable in scholarship. For while the advancement in mechanical time which accompanied the technology of water-powered astraria towers was brought forward by the Chinese culture, all historical accounts utilized in this review, except for one,[35] Joseph Needham's, fall short of exploring (or even mentioning) this coincidence, as it relates to the rapid expansion of clock towers in Europe and Great Britain.

Cultural dissemination may take the form of the written word, or illustrations, carried between geographical locations. As such, the overreaching European view is that the monastic system was directly involved in the transmission of practical concepts – such as water hydraulics – to the secular world that permitted the rate of expansion, and quality of architecture seen in the Late Medieval and later, Renaissance, bell and mechanical clock towers.

These temporal structures reflect a sacred-secular paradigm. They are, at once, structures that encourage (or require) human observation and gathering, while their monumental architecture summons images of vertical constructions as far reaching into time as Egyptian obelisks, and as contemporary as village church steeples. They are underpinned by cosmic time anticipating the rhythm of the heavens in their astrological dials, while presently practical through the addition of clock faces[36] bearing numeric delineations of each hour and (from the late 16th century) minute[37] of linear time. We pilgrimage to see them. We define city centres by their placement, their unmovable location, the quality of the construction, or the charm of their unique features. They appear in Travellers' Guides, books, are the subject of documentaries,[38] and their familiar sound of time has even been known, still now, to introduce the evening news.[39]

Observing the time in front of a clock tower contains an element of theatre that connects us with the tradition of religious celebration. The astronomical clock attributed to the 14th Century Glastonbury monk, Peter Lightfoot (Cox 2008: 183), at the Cathedral for St. Andrew in Wells, Somerset, exemplifies such expression. The cathedral itself is an architectural monument: It is the first (entirely) Gothic-style cathedral in England; its ceilings are supported by a variety of arch works, including the unusual 'scissor arches'; and the structure was, from its inception, designed to be a setting of religious worship for both the ordained, and non-clergical parishioners. A web site is devoted to exploring the history of the space, and visiting options for faith-based, religious, and secular pilgrims. The Wells Clock frontispiece is fitted within the north transept of the church structure, and resides directly above an installation of a wood sculpture of the crucified Christ. Pilgrims are given a sensory performance upon the hour which features as its theme the entwinement of sacred and secular within daily life.

> Its circular dial represents the hours of the day and night, the phases of the moon, and other astronomical [Zodiac] signs; and at the summit is a piece of machinery with figures of knights on horseback, or calvary, which revolve around a centre at the time of striking the hours. At one angle of the transcept is a statue of a seated man, which is connected with the clock by rods,[40] and strikes the hours and quarters with his foot against a bell. This figure, without any intention of punning, is called *Peter Lightfoot.*
>
> (Wells, in Britton and Le Keux 1883: 116)

Perhaps most intriguing, however, is not the mechanization, nor the sensory experience of time; but, rather, the reference to the symbol of Time itself: The clock face is a rare example of a *pre-Copernican*[41] geocentric image of the Earth surrounded by the motion of dials recording the movement of cosmic time. The image of the crescent-moon painted onto the lunar dial ring is encircled by the written words of its creator: *Sphericus archetypum hic monstrat microcosmum* [This spherical globe shows the archetypal microcosm]. Howgrave-Graham (1922: 25–30) has dedicated discussion to the interpretation of this phrase, with emphasis upon the thought that Lightfoot's original phrasing was lost in translation – and in technical transference onto the object. He suggests, among his analysis of the clock's religious and astronomical symbolism, that the characteristics of the word 'microcosm', might be substituted for the idea of 'macrocosm', thereby extending the connotation of the symbolic into the universal domain (Ibid.).

Man the maker

> From the fourteenth century on, the human being, no longer merely a creature, but now also a creator, put himself at the center of the cosmos, and the majesty of the clock was no longer just that of a civil (or religious) counter of the passing years, but took on the precious character of a personal timepiece. Time was adjusted to fit the central position of man: *homo faber.*

> The care devoted to the measurement of time went hand in hand with the gradual mastery by humanity of a world on which a great sun – that of Reason – was never more to set.
>
> (Flechon 2011: 12)

When the door of the time machine was opened onto the centre of a 17th century European city, little would be first recognizable as having endured intact, up to the present. Exceptions to this are architecture; the continuity of our human preoccupation with beautiful objects; and the requirement of accuracy (and later, precision) found in the physics of effective timepieces. We began to wear our clocks as early as the fifteenth century in the form of pocket-sized timepieces (Flechon 2011), designed (still) as much for their aesthetic appearance, as for *observing ourselves observing* the movement of our personal time. By the end of the 16th Century, clocks had moved inside of the Home: Time objects have always been a valuable possession to those who are fortunate in their ability to surround themselves with interesting or beautiful 'things'. Floor clocks take up their own position in rooms where space is not a consideration. Table clocks at their most intricate, resemble sculptures. A simple timepiece, indoors, reminded the household of its status in the process of gentrification. However beautiful, or intricate, or functional, these personal timepieces were, as temporal objects, not shared by All until much closer to *Our Present Time.* (And even, now, the most important clocks in the world are housed in private collections, and in a small group of museums.)

Time's silent neutrality *contained and secularly transformed* in man-made circumstances manufactures the ability to divide[42] human beings through object presence, and subjective utility. Personal time prevails, and yet, still, the temporality inherent in the mechanization process has its limitations: it will not synchronize nature – the solid lands or the liquid seas – nor the rhythms of rural life, despite the clear signal that one may find by retracing one's steps, in the middle of an open field.

Central to an understanding of mechanical time is the thought that, we desire the ability to control the way time accrues and is displayed, with consistency. Consistency requires measurement, equal intervals, demarcated, and (here) advancing, to demonstrate our progress or the completion of a revolutionary cycle, before a new one begins, again. To do so most precisely, requires (an attempt at) removing the remaining imprint of nature-in-time. Mechanized clocks (and, watches) are objects that work *for us*: To transcend human intervention, to reach their potential as machines, they must function independently, self-contained; *they must escape nature, and her natural time.*

What unites a half-millennia of technological progress is the thought that, while the physical parts of temporal mechanization have experienced changes in aesthetics and improvements in design, the earliest physical components – the organs of the clock, such as those Su Sung employed to initiate the temporal motion in his Medieval astraria tower – have remained a constant. The *escapement* still holds back the natural energy of time.[43] It 'escapes' nature by its function as a regulator, releasing and disbursing raw energy at precise intervals to the pendulum (in clocks), or the modern sprung balance (in watches). The organs, in turn, drive the consistency of the horological expression, the ticking over of the gears, the setting of mechanical time in(to) motion. Without an escapement the organs of the clock body would revolve too quickly, and dismember into separate pieces (Flechon 2011: 188). *Time inside the box would cease to exist; the clock would die.*

The pendulum[44] itself requires particular attention, here, as without this innovation, mechanical time would not have progressed further than Su Sung's tower: [Needham suggests] 'that Su Sung's clock was not surpassed as a timekeeper

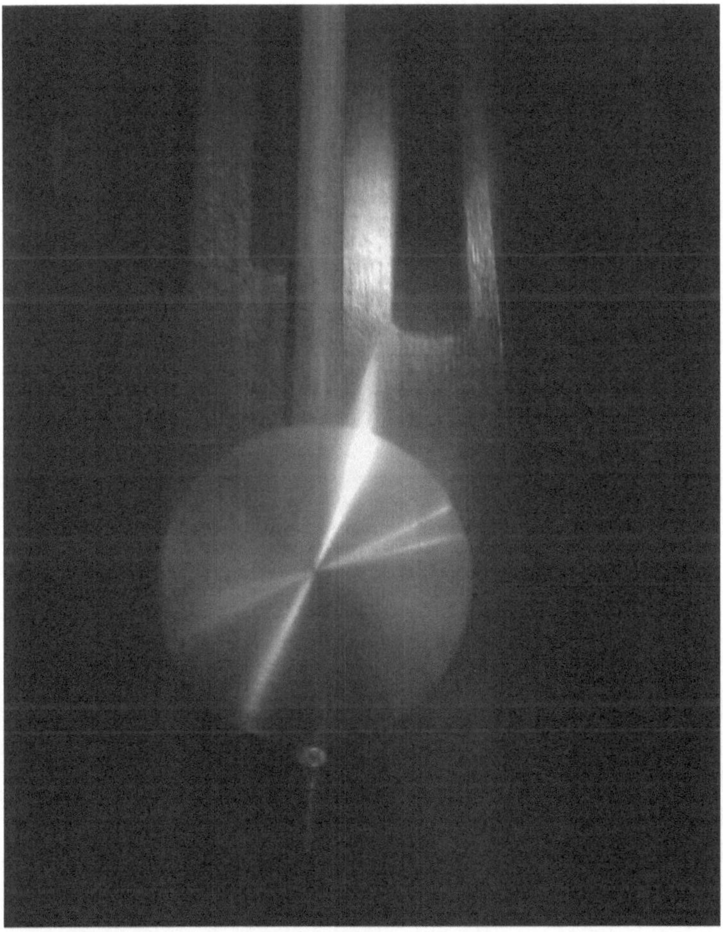

FIGURE 8.7 Cycloidal pendulum with weights.

(Photo: J. A. Swan 2016)

until the invention of the pendulum clock by Christiaan Huygens in the seventeenth century. He may well be right' (Landes 1983: 31). Huygens (1629–1695), a Dutch physicist and astronomer, perfected Galileo's design for a pendulum apparatus, its characteristic polar swing based upon the visual observation of isochronic motion in objects.[45]

The idea that mechanical time is driven to life by continuous polarities of motion is a potent image. The cycloidal pendulum's legacy is the introduction to our scientific ability to mechanically *contain the process of time*, accurately, precisely, and without human intervention.

As creators, we have invented mechanical time, set it into motion, and even, provided housing with a locking door and key. But nature is not easily removed, nor her intentions deterred. A significant issue with *boxing the clock* is that, unless

FIGURE 8.8 Greenwich Observatory Dome, London.

(Photo: J. A. Swan 2015)

the mechanical organs are impervious to the natural world (i.e., sealing the box, or case), they will always remain susceptible to the weathering elements: 'temperature, atmospheric pressure, air humidity, and density' corrupt the accuracy of mechanical time (Flechon 2011: 185).[46]

It was not until the 19th century that the possibility existed for mechanical time to be omnipotent. Gas lighting, and later, electricity, illuminated twenty-four hour days (Frank 2011). *We work into the night, now.* Electric clocks, hung upon walls, and sat upon desks, were re-invented for the managing and monitoring of our life patterns. They emerged in factories, in offices, in schools, public places and spaces (but, not typically, in restaurants or retail shops), replacing the monumental clocks that were neither personal, nor practical to maintain,[47] in the new age of *Universal Time*,[48] and most recently, *Atomic Time*.[49]

Change brings with it unexpected circumstances and conditions. One such example was *the releasing* of the monumental clock towers that remain in city centres to become architectural structures, places of aesthetic pilgrimage for the secular masses. This is not to say, however, that we stopped looking up and into monumental clocks for information about where we are going: While the quality and aesthetic of the temporal object has transformed, passing through any train station or airport, and observing the individuals standing together to check the time of their flight or the arrival of their rail service, one is not without the sense of *observing a pilgrimage*. We wait for time to arrive, to take us to our next destination.

Aesthetic time

Our drive to continue, or, *to keep linear*, ongoing – whether on foot, by ship, railway, or plane, and in our motor vehicles – influences the aesthetics of temporal objects and possessions. Racing sports require the ability to measure the precision of success within time; while soldiers and explorers require temporal devices that will last, through the most extreme human (and natural) conditions. Our modern experience of the collective world is *about timing*, and this connects with the archaic drive, or desire, to carry Time with us, along the journey.

Technology – by the 20th century – has made it entirely possible to mechanically contain the practical facets of this complicated desire within advancing aesthetic forms:

> By the 1880's, watches could display anything that was measurable by mechanical means, function by function grouped together in '*grande complication*' timepieces. This term was to be understood as meaning any function beyond the display of hours, minutes, and seconds. It might be horological: a manual or perpetual calendar, with or without the phases of the Moon, a simple or split-seconds chronograph, or a striking repeater mechanism to which was added, by tradition, a tourbillion and automatic winding. These basic complications could be supplemented with many other

> functions, such as the equation of time, the times at sunrise and sunset, a map of the sky at a certain place . . . universal time, etc. Other complications could be non-horological, such as a thermometer, a barometer, altimeter, depth gauge, or compass.
>
> (Flechon 2011: 16)

The craft of these aesthetics has been established and protected, since the 16th century, when the qualifiable *crafting of time* objects emerged through organized membership in artisan guilds, such as *The Worshipful Company of Clockmakers* in London (est. 1631).[50] Artisan 'seals', such as the *Geneva Seal*[51] (est. 1886), were later established to protect individual watchmakers, ateliers, and their horological craft. Contemporary horological organizations, such as the Geneva-based *Foundation de la Haute Horologerie*[52] (est. 2005), have synthesized an approach to the practical, the historical, and the aesthetic facets of temporal objects, in the 21st century. Such perspective, and continuation, is possible only through the next stage – or change of state – in the process: the collective, symbolic elevation of functional timepieces as works of (wearable) art. *Fine Watchmaking* is now an art: it has experienced a cyclical renaissance, a marriage of superior craft, and (still yet) unparalleled technology:[53]

> We had attained a virtuous circle: the watch was not only considered, studied, and developed with a view to its private, personal function (the watch you have 'on you'), but time measurement was also seen as a reflection, or projection: a mirror of the environment of the wearer.
>
> (Flechon 2011: 16)

The transcendence of functional temporal aesthetics into the symbolic realm, both historic and presently, leaves its earliest residue in portraiture:[54] prominent figures, accompanied or adorned with their personal timepieces. The symbolic treatment of the temporal aesthetic in the visual arts is a curious paradigm. There is a thin distinction between the individualized accessory of time (as in, *the wearing of* or *display of* a timepiece), in portraiture or still life; and the depiction of the clock as a symbolic object, placed deliberately within an environment,[55] or even, (wholly) independent of its practical context.[56] This visual distinction is delineated by historical and cultural factors, as well as artists' style, and specificity of creative *Movement.* The appearance (or continued presence) of the symbol, however, remains atemporal: The symbolic time object does not require the physical or mechanical interventions of human beings to *tell us what time it is.* It is released into time, whole, fixed as an image. Its motion, or 'mechanization', is found in the capacity to visually reflect and connect – from within – the symbolic expression of Time.

Breaking into the visual or symbolic nature of Time is often the remit of artists and writers. In the context of this discussion, it bears mentioning that I was gifted a time object at the beginning of this research process. It was quite unexpected, and yet, the sculpture – designed with the option to be worn as an adornment – encapsulated in one image what I was thinking about writing, at that moment. It resonates here still, and visually reflects much about our human connection within Time. The piece, by Gregory Grozos, a contemporary Greek artist, introduces this chapter: Gregory utilizes what remains of discarded clocks and timepieces – their cases, their organs – to create something new, often portable, sometimes wearable. When I asked him about this process, what *the connection* was for those seeking out (or finding themselves with) this type of sculpture, he replied: 'nostalgia'. It has been some time since we had that conversation, yet, his observation continues to provoke: *What is inside our connection to the Past*, that draws us into dialogues, and (even) disciplines, dedicated to Its research. What is it that are we searching for in Time, and will we ever find that which we seek, *to Know*?

FIGURE 8.9 Omax Spaceman Automatic Watch, 1970s. Collection of Mr. L. Swan. (Photo: J. A. Swan 2016)

Chemical time

The term *pioneer* describes an individual who is present, yet searching ahead; one who initiates the developmental process by seeking understanding, perhaps through seeing into future times; and always, preparing the course or way for others choosing to follow. Ptolemy, Aristotle, Su Sung, Galileo, Copernicus, Huygens, Newton, and Einstein are among the individuals who continue, beyond their own lifetimes, to hold a place within the history of the practical mechanics, the philosophical ideals, and the theoretical nature of Time. Jung, too, pioneered a particular navigation route into our symbolic inner lives, which connects us collectively through Time, from within.

Willard Libby received the 1960 Nobel Prize in Chemistry 'for his method to use carbon-14 for age determination in archaeology, geology, geophysics, and other branches of science.'[57] As the 'pioneer' of the radiocarbon method, the chemical science that transformed the ways in which we study the historic imprint of Human Time,[58] the work of his team is essential to a discussion concerning temporal instruments and objects. Radiocarbon dating investigates the half-life (time) of the carbon-14 isotope (C-14) (produced in the Earth's atmosphere by cosmic radiation and absorbed into all life) to determine the approximate age of an organic object.[59] The imprint of chemical time within organic matter is found through mechanically investigating object life in both cosmic and atomic spheres. *Understanding the space within* an organic life reveals much about its original time. Here, the temporal chemistry of life is initiated through an ending: The determinate age of an organic object (or being) is derived not through an account of *lifespan* itself, but rather, from the approximate time of death, a temporality characterized by decomposition and stasis. In the chemical world life must end for a temporal assignment of life-in-time to be approximately determined.

Technological advancement in science-based material object dating techniques (Aitken 1997) has opened a new window through which to view the vista of the Past. *Who We Have Been*, and significantly, *When We Were There*, may be clarified through the introduction of the physical sciences into time keeping. Perhaps the most significant collective shift in our contemporary thinking about human timelines – particularly, in European culture – involves the introduction of radiocarbon technology and dendrochronological calibration (Renfrew 1973) to disprove the linear-driven evolutionary concept of the *invasion* (or diffusion) *hypothesis*.

> Most of us [born before the 1970s] were brought up to believe, for instance, that the Pyramids of Egypt are the oldest stone-built monuments in the word, and that the first temples built by man were situated in the Near East, in that fertile land of Mesopotamia . . . The knowledge of working in copper and bronze, like monumental architecture and many other skills, would then have been acquired by less advanced inhabitants

FIGURE 8.10 The Antikythera Mechanism, c. 200 BC.
(Athens: National Archaeological Museum)

> of surrounding areas, and gradually have been diffused over much of Europe and the rest of the Old World . . . It comes as a shock, then, to learn that all of this is wrong. The megalithic chamber tombs of western Europe are now dated earlier than the Pyramids – indeed they rank as the earliest stone monuments in the world . . . And Stonehenge was, it seems, completed, and the rich bronze age of Britain well under way, before the Mycenaean civilization of Greece even began . . . The traditional view of prehistory is now contradicted at every point.
>
> (Renfrew 2011: 1–2)

The radiocarbon studies of the 1960s and 1970s produced material evidence from *sites and objects* in the Ancient Near East, Europe, and the British Isles that transformed the linear model of cultural flow across geographic distances (Ibid.). Perceptions of human time now hold the recognition that the scientifically identifiable *temporal presence* of a given stone, artefact, or metallic object *at a particular point in time* solidifies and relativizes the whole expression of the architectural site, within a measurable region of Historic (Human) timeline. The thought that the emergences of organized cultural sites – particularly in British regions and the Americas far removed geographically from the Near East – developed independently, in their own timeline while within the collective timeline *supported a different, non-linear quality of cultural motion.*

Object synchronicity

The thought persists, that the life of objects – and the *life inside* those objects – has given science the ability to re-calibrate our understanding of Time, is not without coincidence: It is at this intersection of objects – here, ancient artefacts – and the technological advancement of culture, that coincidences occur, and the story of non-linear object resonance begins to take meaningful form, beyond what we have come to understand as *inanimate* matter in our material world. Two instances of *object synchronicity* are helpful to an understanding of the significant effect temporal objects have had upon our ability to relativize (to qualify, and to quantify) Past Times.

The National Archeological Museum in Athens houses an extensive collection of ancient artifacts, including a curious object that was found, among others, in 1901, by sponge divers combing the (then) undiscovered shipwreck of an Ancient Roman vessel off the coast of the island Antikythera in the Aegean Sea. Among the coins, statues, and household goods, a large, corroded metal and wood object was brought to shore and taken into storage. It remained dormant, until the 1950s, when early x-ray technology established that the object was composed of 'gear wheels, dials, and hands' (Flechon 2011: 66), and that '[the] gear wheels were similar to the first astronomical clocks of the Renaissance' (Ibid.: 66–67).

Details of the *Antikythera Mechanism Project*[60] suggest that this object continues presently to stimulate temporally sensitive research in multi-disciplinary academic fields. Computer Tomography (CT) technology, since the 21st century, has expanded the study, permitting detailed scanning and virtual cataloguing of the artifact's fragments. Though data remains unclear – at this stage of research, more than one century along – and unable to verify the popular theory that the artifact is perhaps the oldest example of a mechanized computation device, a *cosmic computer*.

The temporality inherent in the Antikythera Mechanism's libido remains, however, potent: Its existence provides physical evidence that the Ancient Greeks were capable of translating their theories into practical time objects, in a far more sophisticated arrangement than previously considered. Furthermore, that the object itself continues to intrigue an international community of scholars, scientists, and technology professionals, suggests not only a re-awakening of dormant object libido – since the 1950s – but also, the capacity for object transcendence, continuation within a new, atemporal lifetime beyond its linear birth into material existence.

Earlier, the conceptualization of cosmic calendars was presented in the context of the written word and emerging Ancient cultures. Returning here to a different type of artifact, brought into material life millennia before the advent of organized record-keeping, Ruggles (2005: 5) states that the earliest calendars – of stone, bone, and antler – consisted, 'of a series of marks' depicting what is thought to be consistent with a pattern in observable cosmic time. Under microscopic analysis, the archeologist Alexander Marshack hypothesized that 'the marks were not a simple decoration, but carefully accumulated, often using different tools and techniques, over a period of time' (Ibid.: 6).

What is perhaps most curious is that, like the Antikythera mechanism, these Paleolithic artifacts re-emerged in (seemingly) random circumstances that facilitated the active re-engagement of object libido with a precise human interest: the seeking of knowledge about the history of Human Time. The archaeologist undertaking the examination of these Palaeolithic calendars was, initially, following his research work for a NASA project when he encountered the artifacts in store at the Musee des Antiques Nationales, where they had been lying dormant for the second time, since their initial archeological unearthing.[61] The radiocarbon dates of these bones, together with the still-controversial theory[62] of their distinct markings, established the possibility that human beings had been looking up and into the patterns of the sky, and recording those observations, far longer – for twenty thousand years – than any scholarship had previously anticipated.

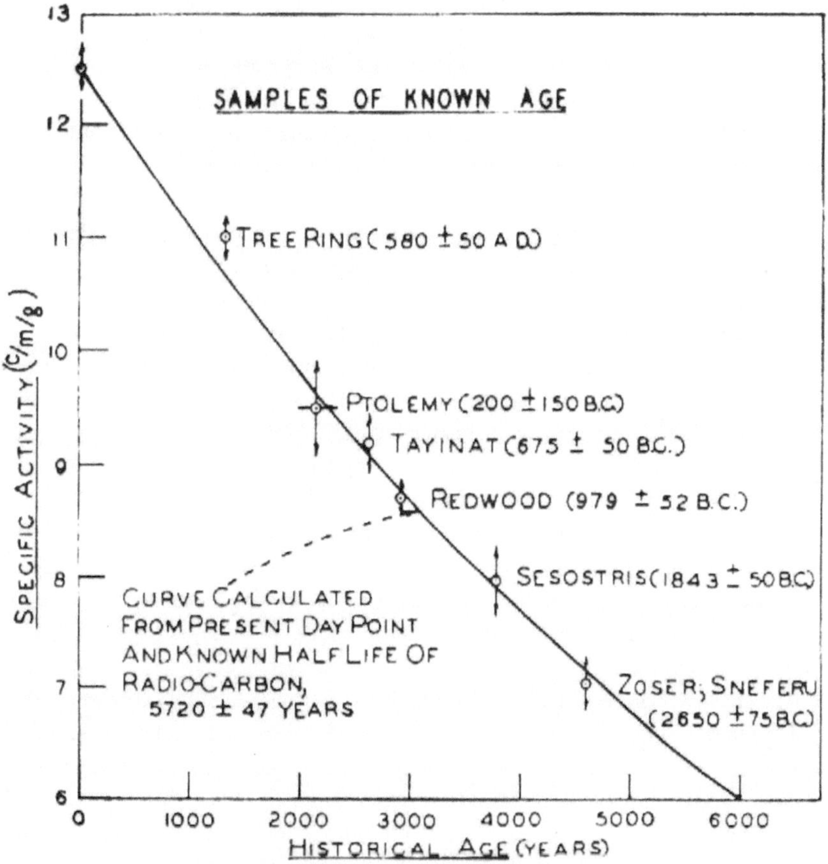

FIGURE 8.11 1940s hand-drawn scientific temporal graph.

In, Arnold, J. R. and Libby, W. F. (1949). 'Age Determinations by Radiocarbon Content: Checks with Samples of Known Age'. Science, v.110, pages unlisted.

A natural re-birth

> Trees, as it is well-known, have played a large part in religion and in mythology from the remotest times. Typical of the trees found in myth is the tree of paradise, or tree of life . . . In this world of images the Cross is the Tree of Life and at the same time a Tree of Death, a coffin. Just as the myths tell us that we were descended from trees, so there were burial customs in which people were buried in hollowed tree-trunks, whence the German *Totenbaum*, 'tree of death', for coffin . . . If we remember that the tree is predominately a mother symbol, then the meaning of this mode of burial becomes clear. The dead are delivered back to the mother for rebirth.
>
> (Jung CW 5: 349)

The temporal death of matter is most essential to the chemistry of Time. Trees – and wood objects that originated from felled trees – are the source of natural time in the mechanics of radiocarbon technology, and its linear, scientific advancement. Dendrochronology and its variants examines the patterns of an earth-bound natural calendar, the concentric circles defining the lifecycle of a tree. This organic calendar provides a complement (and methodological supplement) to the accuracy of the radiocarbon dating process.

Developing an organized system for the advent of early radiocarbon technology required establishing a base of *known-in-time* objects, illustrated by Libby's *Curve of Knowns*, from which to compare the chemical measurement (and accuracy of the measurement) recordings of the carbon isotope presence.

Libby's early radiocarbon investigation (Arnold and Libby 1949) provides a different imagining of temporal objects in the psychic realm, and their capacity to express synchronistic features characteristic of the psychoid expression of Time. This first-published study on radiocarbon dating (Ibid.) utilized samples from tree specimens and five wood artefacts, carved and tooled by human hand. The trees accounted for a sample cross-section ring of felled fir, living between 530 AD and 623 AD, and a heartwood section of the *Centennia Stump*, 'one of the largest redwood trees ever cut . . . the carbon in the central wood was deposited there about 3,000 years ago, although the tree itself was cut just a few years ago' (Libby 1960: 602); while four of the five[63] wood objects were funerary in origin, found in Ancient Egyptian tombs, dated from Early, Middle and Late Dynastic Periods.[64]

Egyptian religion required the corpse of the deceased of be mummified to preserve the soul, or *Ba*, intact for its journey by solar ship into the Afterlife (Zakbar 1968). *The dead are delivered back to their mother for rebirth.* This mythological interpretation of temporal (and object) continuity in Death is reflected in the thought that Egyptians' tomb furnishings were often purpose-built objects – such as beds or boats – for continuity of life in the Afterlife-time. Yet, in linear time, these objects remain visually dormant until found within a container-for-containment of the *Lives of the Dead*, excavated, re-activated, and repurposed within time. That these wooden *afterlife* objects could remain at all (not merely, for five millennia), and that they carried their chemical – as well as cultural – information through providing an underpinning for research that resulted in the re-calibration of our Human history, is remarkable.

Looking into the sky at night brings the visions of the ancient cosmological world into focus. The image and mythology of Senusert III is potent: he appears on the deck of his solar ship's maiden voyage through the cosmic sea. *A chip in the smooth decking boards reveals a temporal scar that has transcended linear time.* We will never know if Senusert – or even, *his Ba* – has, indeed, reached the desired destination (or encountered his destiny). But the observation prevails that the objects of his death, imbued with the regal essence of cosmic time, have been brought forth under new circumstances and developing conditions, and as such, continue now to live again.

A synchronos path

In view of an ending (or new beginning) to this path along the history of temporal life in objects, it would seem appropriate to reflect upon the curious expressions of time encountered along this journey. First, much has been said about the concept of enigmatic entanglement: The thought that, as we *continue* in our collective human development, our relationship with life matter continues to refine; and, that this feature of culture is observable through the linear progression of our understanding of Time, found in the material technology of temporal sites and instruments, and constructed through the abstracted theories of the physical sciences. Underpinning this paradigm is the observation that, without the innovativeness characterized by the creative spirit, and the animation of psyche life, much of what we understand about Time – and our lives within times – would be without *direction* or *purpose*, two prominent cognitive features of human existence brought forth through perceptions of meaning. There are, however, limitations to our human capacity for observation in the material world. We are subjective beings, bound to space and place *within Time*. Our understanding of lifetime is found in incremental moments of clarity as we learn who it is we are becoming, and how we are making our journey – our circle – whole.

In this story of temporal sites, instruments, and objects a conclusion is reached by returning to the *state of being*, from object origins in the material world as our cultural heritage and possessions, to objects' libido as living 'things'. Through the process of being *created into a lifetime*, an object's libido – activated by contact with humankind – innately possesses the capacity for change, or even, a transformative state, and as such holds the possibility, or quality, to physically and metaphorically reflect unto us our own history within Time.

There is the thought, in archaeology and anthropology, that material objects are capable of aligning, or synchronizing, our understanding of Human Time. It remains true that much of what we understand about artefacts from the Past –

beyond the assignment of chemical time – is established through our present human observations, our *projections* of object lifetime, and through information *left behind* in written records. We have been writing in time for five millennia, and our hand-written (and later, printed) recorded observations of lifetimes – in calendars, documents, letters, diaries, and books – provide us with a structure or scaffold upon which to organize historical life across cultures, through millennia.

Nevertheless, however experienced, mechanical precision and linear accuracy remain unchallenged polarities of human observation and Natural time. Our understanding continues within a circle, a cycle. We grow into this understanding through revolutions of a universal cycle, the learning and re-learning of a *life-in-time* expanding (and even remaining beyond) its libido. What we come to know changes across time, and us with it. This is how we grow. Underpinning this thought is the observation that as developing beings what we understand in a moment, or hour, or day, or even years, is sometimes quite different from what is accurately or precisely the Truth, in Time.

Conclusion: the journey

The A303 road is part of the English network of travel routes that connects greater London to Land's End. Like many rural roads, the carriage way narrows to a single lane in each direction, through a length that joins the ancient town of Amesbury and the parish village of Winterbourne Stoke. A traveller familiar to this journey might mention a congestion, perhaps. To the uninitiated, this observation is

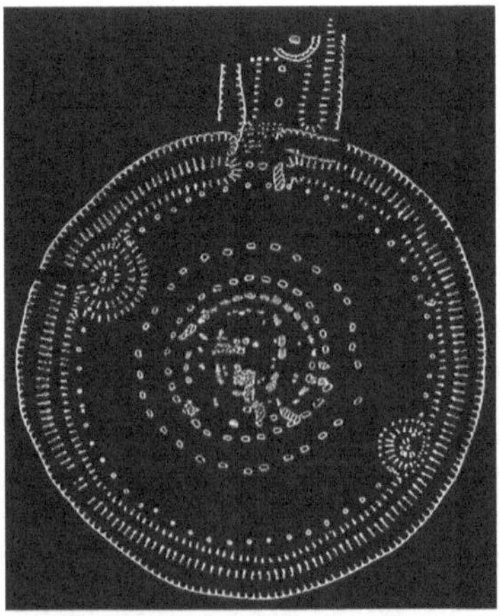

FIGURE 8.12 Solarized aerial view of UNESCO Stonehenge Archaeastronomical site, 1970s.

Inspired by illustrated surveys of the Department of the Environment (UK), and 1970s illustrations by Evan Hadingham (J. A. Swan 2016).

incongruous: a surrounding view of undeveloped greenness, a pathway driven through farmland, fields of unaffected millennia. Descending a hill, the traffic slows, moving, now boot to bonnet, as one would walk the way there. Yet, there are no dramas of collision, no works in progress. Only the silent sky and the sounds of the natural earth. Ahead, a ring of stones, standing. They are still observing.[65]

Notes

1 Ibid. 4

2 'We construe that the past, the present and the future are *not sequentially arranged* in the psyche . . . Specifically, if we consider the possibility that in the psyche's linear time's demarcations all coexist in a relative state, then everything – and at the same time nothing – is indeed *simultaneous* and *contemporaneous*' (Yiassemides 2014: 37).

3 (1933: 95–116).

4 Jung, C.G. (1911–1912/1952)/CW5: 425.

5 Referring to the libido and time: 'It is not simply that the opposition is inherent in both and that they thus could be conceptualized as comparable, or even equivalent energies or concepts. In that case opposition would have been the force for which Jung attempted to find a symbolic expression via time and libido. Rather, this quote indicates that time is represented *by* the libido' (Yiassemides 2014: 3).

6 Armstrong (1996: 27–34, 51–52, 98–99); Cioran (1970), Eliade (1987: 134, 207).

7 Griffin and Tyrrell (2011: 33, 35) suggest that temporality in conscious thinking emerged during the 'Big Bang' of the Upper Paleolithic era giving rise to abstracted thought, meaning-making, and a quality of 'mysticism' or 'intuition' that facilitated the possibility to 'escape space and time and become aware of the network of relationships that hold the universe together.'

8 'Archaeoastronomy is the study of beliefs and practices relating to the sky in the past, especially in pre-history, and the uses to which people's knowledge of the skies was put.' Ruggles, C., accessed 7 July 2015 at www.cliveruggles.net. For examples of sites and investigations, see Ruggles, C. (ed.) (2011, 2015). See also www.archaeoastronomy.org (SEAC); www2.archaeoastronomy.org (ISAAC); and www2.astronomicalheritage.net (UNESCO's Astronomy and World Heritage Initiative).

9 The use of the generalized terminology 'thing' appears formally in archaeology and anthropology, and when referring to scholarship associated with Malafouris' and Renfrew's *Material Engagement Theory* (MET). See Appaduri (1986), Malafouris and Renfrew (eds) (2010).

10 'The earliest indications of the use of symbolic notation to represent or visualize an astronomical cycle comes from the Upper Paleolithic period . . . engravings were made on small portable objects such as stones and pieces of bone and antler. Thousands of examples are known. A number of these are in the form of a series of marks, and several were subjected to meticulous microscopic analysis by the American researcher Alexander Marshack [1918–2004]. He concluded that the marks were not a simple decoration, but carefully accumulated, often using different tools and techniques, over a period of time' (Ruggles 2005: 5–6).

11 Ibid.

12 'Since the *unus mundus* represents the unified reality of the psychoid world it can be viewed as the container of the conscious and unconscious parts of the psyche alike and, consequently, of their corresponding temporal realities' (Yiassemides 2014: 71).

13 The earliest cohesive systems of writing emerged in the Ancient Near East (Mesopotamia c. 3300 and Egypt c. 3200).

14 Lord Colin Renfrew was the first to use this terminology evincing the dramatic changes in the field of archeology following the discovery of carbon dating technology, in the 1950s by Willard Libby and his research team (Renfrew 1973).
15 'It was possible – as evidence from the Middle Kingdom site of Illahum bears out – that a temple might schedule its festivals according to a purely lunar calendar. Indeed, it appears to have done exactly that for more than a millennium (from ca. 3100 BC to 1800 BC). But, still, the temples were forced to date these festivals in its records according to the accepted [solar] civil calendar' (Spalinger 1999, in Lippincott 1999: 268).
16 www.etymonline.com, accessed 7 July 2015. '[L]iterally, 'one that discerns or examines' from gignoskein 'come to know' (Gnostic, adj).'
17 www.etymonline.com, accessed 7 July 2015.
18 1327 AD.
19 'In around 529, St. Benedict set the rule that fixed the hours of prayer, study, and manual work which rapidly developed into the canonical hours used in certain countries until the middle of the sixteenth century. In addition to consulting the sundials, the rule of St. Benedict required monks to identify the moment of midnight, the time for confession, and this meant using lunar dials and clepsydras. At the end of the fifteenth century, the *De cursu stellarum* by Gregory Tours laid down a group of principles, based on nocturnal observation, that enabled monks to know when it was time for their nocturnal prayers.' (Flechon 2011: 107).
20 In this way, it shares the symbolism of the sealed-gear mechanism of table clocks from the 16th century forward, and by extension the decorative pocket and wrist watch.
21 Water clocks appeared in Ancient Egypt as early as 1500 BCE, with artifacts dating to the reign of Amenhotep III, 1450–1380 BCE; Written evidence exists of their use in Mesopotamian (Babylonian) culture by 600 BCE. (Flechon 2011: 35, 55, 433).
22 'The final classical citation of such a time-telling device can be found in a text written sometime around 520 AD by the late Roman author Cassiodorus (Instititionum, I, xxx, 5).' (Lippincott 1999: 132).
23 The earliest – also, the most basic of design, generally – models, a *simple outflow* device, utilize a container (carved stone or metal bowl) with incremental markings inscribed on the interior walls. A regulated drip or flow of water diminishes over a period, thus permitting time to be measured by either the water meniscus, or an object placed within the water, as it reaches (or, sinks to) a particular height at a given moment. The Ancient Greeks developed an advanced system of gear mechanism, an *inflow* device, wherein liquid was contained externally – and introduced to the system incrementally – improving the quality of water flow into the time-keeping mechanism, thus extending the possibility to measure time more accurately, and across a longer period (Lippincott 1999: 124–25, 132; Flechon 2011: 55, 123).
24 Greek version of the water clock: 'water thief', from Greek *klepsydra* 'to steal, to hide' and 'water'.
25 Anaphoric clocks are considered the earliest form of mechanized time-keeping devices. They initially appeared in Ancient Greece (c. 200 BCE). Like the Mesopotamian planisphere, and the Arabic astrolabe, they reproduced a two-dimensional projection of the cosmic cycle. The water-driven mechanism supported movement of the two-dimensional dial, which tracked the pattern of solar and lunar movement across the increments of days, and the rising and setting of various celestial bodies within the cosmic cycle (Lippincott 1999: 138–39).
26 A generic term used to describe any time-keeping device, from the most primitive sun clock, to the most advanced mechanized gear structured apparatus (Flechon 2011: 104; Lippincott 1999: 132).
27 Mechanical device that displays the motion of the cosmic bodies, used for study of heavenly patterns, rather than precision, in time-keeping (Landes 1983: 29).

28 Arabic water clocks of the Middle Ages are technologically unsurpassed (Flechon 2011: 85–87).
29 See Landes (1983: 392 n.10) regarding the similarities between Islamic and Chinese water clocks.
30 'An instrument provided with circles representing, in particular, the equator, the ecliptic, the tropics of Cancer and Capricorn and the polar circles. As the earth is at the center of all of these circles, an armillary sphere can be used to tell the time.' (Flechon 2011: 430).
31 The modern escapement 'counts the beats (or pulse) of the clock by blocking and releasing the wheel chain' at a specific rhythm (Landes 1983: 9).
32 An excerpt from Su Sung's Memorial to The Emperor Che Tsung. Joseph Needham *et al.* (1986) provide a full translation of Su Sung's horological treatise, and accompanying work on horological technology.
33 Title of a 13th century mechanical illustration by the artist, Villard de Honnecout (Flechon 2011: 110).
34 'The list of cities with public clocks begins with Orvieto, Italy, in 1307, followed by Modena in 1309, Parma in 1317, Ragusa in 1322, Milan in 1336 and Padua in 1344. By mid-century, mechanical clocks began to appear outside of Italy: Windsor Castle gained a clock in 1353, Avignon in 1353, Prague in 1354, Regensburg in 1358. By the beginning of the fifteenth century, public clocks had evolved into the standard in even smaller settlements. The town clock became a matter of both civic need and civic pride (Frank 2011: 72).
35 See Needham et al. (1986), and reference to Needham's life work on Chinese science and technology in Landes (1983: 17–36, 37–52).
36 Clock 'faces' are Northern European in their origin, and appeared during the early 15th Century. Prior to this, clock towers were functional containers for bells, topped with architectural variants of a steeple, turret, or ornamental spire (Flechon 2011: 132). Bell Towers appeared in both secular and sacred settings, accompanying Town Halls and cathedrals. This image translates into the sacred architecture of church steeples, housing bells, calling the pious to prayer.
37 'It was the minute hand, in fact, that would propel time into a new dimension. The minute was a small but workable unit of time . . . It announced a new kind of time that would govern the home, the workplace, and most important, the laboratory' (Frank 2011: 91).
38 For example, the Prague Astronomical Clock (the *Prague Orloj*) celebrated its 600th anniversary in 2010, with a documentary of the celebrations.
39 The BBC News 4 has been using the Great Bell (*Big Ben*) in the Clock Tower (from 2012, the Elizabeth Tower) at the Houses of Parliament, since 1924. The tolling is heard live each day before news broadcasts at 6pm and midnight through a microphone link directly to BBC headquarters at Broadcasting House, London. See www.bbc.co.uk/ahistoryoftheworld/objects. Perhaps the most iconic secular image associated with London, the clock itself – and by extension its tower – is often referred to (particularly by those outside of the UK) as a unified object, rather than a container for 'Big Ben'.
40 The Wells Clock incorporates the oldest jaquemart, or bellstriker, a figure also known by some as Jack Blandifer (Jagger 1977: 64).
41 Nicolaus Copernicus (1473–1543), Polish astronomer and mathematician, author of *On the Revolutions of the Celestial Spheres* (1543), holds a particular resonance in this discussion of the heliocentric model of the psyche: It is his heliocentric model of our solar system – that while provocative, in Renaissance Europe – has proven, as the Copernican revolution, to become an essential scientific understanding. Wallace (1987: ix–x) remarks: 'Although Copernicus' work was predominately scientific, its consequences, of course, extend far beyond science, inspiring philosophies, causing problems

for religion, and producing powerful metaphors like that of man's "removal from the center" of the cosmos . . . the modern world is seen as a product essentially of changes originating within the self-contained activity called "science" . . . This is especially so when we remember that the notion of science as a self-contained and self-regulating long term process itself originated only in the seventeenth century, at the same time that the Copernican revolution was taking hold.'

42 Time also has the ability to unite, in the form of Revolutions.

43 Derived from weights or springs.

44 This passage concerns the cycloidal pendulum, characterized by an oscillating, or isochronic, pattern, that is standard apparatus employed in pendulum-driven clockworks, to present day.

The perhaps more familiar pendulum, by title, if nothing else, is the one designed by the French physicist, Jean Bernard Foucault, in 1851. 'Foucault's pendulum experiment provides a clear and convincing demonstration of the rotation of the Earth . . . The observed result is that the plane of swing of a pendulum changes with time. In the Northern Hemisphere, the plane of swing moves in the same sense as a clock which is face up on the ground; in the Southern Hemisphere, it moves in the opposite sense.' (Sommerville 1972: 40).

45 Galileo Galilei (1564–1642) is thought to have been inspired to design the pendulum apparatus through observing the regular movement of a lamp hanging from the ceiling in the cathedral at Pisa, during mass (Flechon 2011: 184). Contemporary science has teased apart the connotation of Galileo's observation, and design, reflecting that, 'Because secondary perception can be so heavily influenced by belief and social setting, it can lead to error, as in the case of Galileo. His theory . . . turned out to be wrong. Ideal circular pendulums with equal length, but unequal amplitudes of oscillation, do not stay in synchronization. (Ideal cycloidal pendulums are required for that)' (Lomas 2005: 298). It is worth noting that creative ideas often continue to grow beyond their inception, and as such, are susceptible to interpretation, and change: Huygens took direct inspiration from Galileo's ideas on the pendulum, and his 1659 design and model reflects the presence of a 'cycloidal' pendulum, the accuracy of which is unsurpassed, and continues to be found in pendulum clocks today (Flechon 2011: 183).

46 These conditions were particularly true of temporal-navigational marine clocks fitted for nautical voyages, that grew in prominence along with trade routes and geographic explorations, from the 17th century forward.

47 In 2010, after more than six centuries of being hand-wound by clock-keepers and horologists, The Great Clock at Wells Cathedral (est. 1392 A.D.) was fitted with an electric motor. The motor automates the mechanics, for ease of service, and lack of individuals available to perform the thrice-weekly winding process. See www.bbc.co.uk/news/uk-england-somerset-11040583.

48 Coordinated Universal Time (est. 1967) is the standard for civil time across the world. Initially known as simply *Universal Time* (est. 1884 at the International Meridian Conference in Washington, D.C.), the system is based upon a global-standardized 24-hour time zone model, wherein Greenwich Mean Time (GMT) is the Prime Meridian. Presently, CUT is composed of two readings: Universal Time (UT 1), or astronomical time, measuring the Earth's rotation; and International Atomic Time (TAI) (Royal Museums Greenwich 2012; tycho.usno.navy.mil).

49 International Atomic Time (TAI) is achieved through scientifically monitoring the oscillation frequency of the Cesium-133 atom: Atomic clocks measure the number of oscillations per second unit of time (more than 9 million) to count time passing. The International System (SI) aligned the 'second' unit of measurement with Atomic time, in 1967. Currently, 69 laboratories worldwide (more than 400 atomic clocks in total) provide ongoing data, which is synchronized using satellite technology. This technology supports the temporal life of devices that function with hyper-accuracy: mobile

phones, computers, GPS systems, and military hardware (Frank 2011; tycho.usno.navy.mil).

50 See www.clockmakers.org, official site.

51 The seal has become a matter for public (and governmental) debate, however, as its initial intention was to protect craftsmen and the quality of their work (only) within the canton of Geneva; The use of artisans seals in Switzerland also comes under the jurisdiction of the COSC (Control Officiel Suisse des Chronometres), that tests and verifies the scientific precision of timepieces.

52 See www.hautehorlogerie.org, official site. The Foundation's multidisciplinary approach provides educational training, academic resources, and cultural experiences for scholars, watchmakers and enthusiasts. Foundation Chairman, Franco Cologni, and Committee Member, Dominique Flechon, collaborated in *The Mastery of Time* (2011).

53 'The battery, a source of environmental pollution represents one of the major drawbacks of this technology. Solutions have been found by combining mechanics and electronics. Nevertheless, faced with a module that they considered vulgarized and that inspired no emotions, aesthetes and collectors agitated for a return to tradition. From [mid-20th century] on, watch making was to blossom into a new system of multiple branches comprising watches that might be mechanical or electronic, but all corresponding to different approaches to the perception of time' (Cologni, in Flechon 2011: 17).

54 Two early examples include: Manzuoli, T. (c. 1558). 'Portrait of Man Holding a Watch', and Hals, F. (1643). 'Portrait of a Man Holding a Watch'.

55 E.g., de Pereda, A. (1632–1636). 'Allegory of Vanity'; Dali, S. (1931). 'The Persistence of Memory'; Munch, E. (1940–1943). 'Self-Portrait Between the Clock and the Bed'; Chagall, M. (1930–1939). 'Time is a River Without Banks'.

56 E.g., Duchamp, M. (1915–1923). 'The Bride Stripped Bare By Her Bachelors, Even The Large Glass'; Gonzales-Torres, F. (1987–1990). 'Untitled (Perfect Lovers)'; Marclay, C. (2011). 'The Clock'.

57 www.nobelprize.org/nobel_prizes/chemistry/laureates/1960/libby-facts.html, accessed 7 July 2015.

58 In the seven decades since Libby and his team begun their work, the Carbon-14 dating method has been complemented – and in certain instances surpassed (in accuracy) – by the continuation of study: We are interested, and committed scientifically, to characterizing and organizing past human times (and the objects of these times) as accurately, as possible; Technological advancement of radioactive (here, cosmogenic) dating techniques has permitted accurate dating of inorganic matter, such as megaliths, rock formations, and metals. Chemical time-determination techniques are now complemented by computer-driven technology, used to digitally examine and analyse objects and on-site excavations. The organic (chemical) feature of time-telling has also merged with technology, forming new temporal instruments that incorporate an all-in-one-chemical time recording device (Aitken 1997).

59 While an organic substance – most commonly wood or plant materials, human and animal remains – is alive, the presence of C-14 exists as a constant; However, upon death (whether human or plant matter), the constant percentage of the isotope begins to 'decay', at a measurable half-life (5,730 years). This technology remains limited: first, that only organic matter may be investigated using the Carbon 14 method; second, that the older an object is within time, the more difficult it is to maintain accuracy in discovery; third, an object may experience 'contamination' from being proximate to other materials, this is especially true in the case of sites where layers of time have accumulated, and geographical and climate changes have impacted the integrity of the site. Since the 1970s, C-14 dating has become more accurate in its ability to tell time by individually 'counting' the carbon isotopes, through the technology of Accelerator-based Mass Spectrometry (AMS). Often, radio-carbon dating is combined with a variant

of dendrochronological (tree or growth ring) dating. See Laboratory of Tree-Ring Research, www.lttr.arizona.edu.

60 See www.antikythera-mechanism.gr.

61 Marshack's (1972: 35, 44–45) *Roots of Civilization: The Cognitive Beginnings of Man's First Art, Symbol and Notation* recounts this 1962 discovery. Particular to his micro-photographic study of artifacts are the *Lartet Bone* unearthed in 1865 in the Vezere Valley (near Lascaux cave site), and the *Blanchard Bone* unearthed in 1915, in the Dordogne, among other samples.

62 Marshack's theory of cyclical marking based on his interpretation of micro-photographic evidence continues to be considered in doubt by members of the archeological (Paleolithic) community (Rosenberg 2007: n.11).

63 The fifth sample was a wood section of floorboard in the (burnt-out) central room of a Syro-Hittite 'hilani', or palace, in the city of Tayinat (present day Turkey) (Arnold and Libby 1949).

64 These include wood samples from: a mummiform coffin, Ptolemaic period (332–330 B.C.); from the Middle Kingdom, a piece of decking board from the Chicago funerary barge of Senusert III (1843 B.C.); and two samples from the Early Dynastic period, from the tombs of Sneferu (4575 B.C.) and Zoser (4650 B.C.) (Arnold and Libby 1949).

65 In the previous twenty-five years there have been fifty-one separate proposals made to change the commuting pattern of the A303 rural road passing the Archaeastronomical UNESCO World Heritage Site, *Stonehenge*. A new visitors centre was erected in 2013, five millennia after the placement of the Heel stone, to accommodate for the increasing annual pilgrimage of 1.3 million individuals. Our initiated, collective engagement with cosmic time continues. See www.bbc.co.uk/news/uk-england-wiltshire-30529065 and www.unesco.org.uk/UserFiles/Files/UKNC_Stonehenge_042006.pdf.

References

Aitken, M. J. (1997). *Science-Based Dating in Archaeology*, London, Routledge

Appadurai, A. (ed.) (1986). *The Social Life of Things*, Cambridge, UK, Cambridge University Press

Armstrong, K. (1996). *In the Beginning: A New Interpretation of Genesis*, New York, Ballantine

Arnold, J. R. and Libby, W. F. (1949). Age Determinations by Radiocarbon Content: Checks with Samples of Known Age. *Science* 110

Britton, J. and Le Keux, J. (1883; 2013 ebook ed.). *A Dictionary for the Architecture and Archaeology of the Middle Ages*, London, James Moyes

Brown, P. (1976). *Megaliths, Myths and Men*, New York, Dover

Cioran, E. (1970). *The Fall into Time*, trans. R. Howard, New York, Quadrangle

Cologni, F. (2011). Introduction, in Flechon, D. (2011). *The Mastery of Time*, Paris, Flammarion

Cox, J. C. (2008). *English Church Fittings, Furniture and Accessories*, London, JM Classic Editions

Donald, M. (1991). *Origins of the Human Mind: Three Stages in the Evolution of Culture and Cognition*, Cambridge, Massachusetts, Harvard University Press

Eco, U. (1999). Times, in Lippincott, K., *The Story of Time*, London, Merrell Holberton

Eco, U. (2001). *The Name of the Rose*, trans. W. Weaver, London, The Folio Society

Eliade, M. (1987). *The Sacred and the Profane*, New York, Harvest/Harcourt

Flechon, D. (2011). *The Mastery of Time*, Paris, Flammarion

Frank, A. (2011). *About Time*, London, Oneworld

Griffin, J. and Tyrrell, I. (2011). *Godhead: The Brain's Big Bang*, Chalvington, UK, Human Givens

Haldane, C. (1984). *The Dashur Boats*, College Station, Texas, Texas A & M University

Hodder, I. (2012). *Entangled: An Archaeology of the Relationships Between Humans and Things*, Chichester, UK, Wiley-Blackwell

Howgrave-Graham, R. P. (1922). *Peter Lightfoot – Monk of Glastonbury and the Old Clock at Wells*, Glastonbury, UK, Avalon

Jagger, C. (1977). *The World's Great Clocks and Watches*, London, Hamlyn

Jung, C. G. (1911–12/1952). *Symbols of Transformation*, in Sir H. Read, M. Fordham and G. Adler (eds), W. McGuire (executive ed.), trans. R. F. C. Hull, *The Collected Works of C. G. Jung* [hereafter CW] Vol. 5, London, Routledge & Kegan Paul

——. (1921/1971). *Psychological Types*, in Sir H. Read, M. Fordham and G. Adler (eds), W. McGuire (executive ed.), trans. R. F. C. Hull, CW Vol. 6, London, Routledge & Kegan Paul

——. (1955–56/1963). *Mysterium Coniunctionis*, in Sir H. Read, M. Fordham and G. Adler (eds), W. McGuire (executive ed.), trans. R. F. C. Hull, CW Vol. 14, London, Routledge & Kegan Paul

——. (1933 [1955]). The Stages of Life, in *Modern Man in Search of a Soul*, London, Harcourt Brace Jovanovich

——. (1995). *Memories, Dreams, Reflections*, London, Fontana/Harper Collins

Landes, D. (1983). *Revolution in Time*, Cambridge, Massachusetts, Harvard University Press

Libby, W. (1960). 'Radiocarbon Dating', Nobel Lecture. (595–610), www.nobelprize.org/nobel_prizes/chemistry/laureates/1960/libby-lecture.pdf, accessed 7 July 2015

Lippincott, K. (1999). *The Story of Time*, London, Merrell Holberton

Lomas, D. (2005). Degree of Influence on Perception of Belief and Social Setting: Its Relevance to Understanding Pendulum Motion, in M. Matthews, C. Gauld and A. Stinner, *The Pendulum: Scientific, Historical, Philosophical and Educational Perspectives*, Dordrecht, Nebraska, Springer

Malafouris, L. and Renfrew C. (2010). The Cognitive Life of Things: Archaeology, Material Engagement and the Extended Mind, in *The Cognitive Life of Things: Recasting the Boundaries of the Mind*, L. Malafouris and C. Renfrew (eds), Cambridge, UK, McDonald

Malafouris, L. and Renfrew, C. (eds). (2010). *The Cognitive Life of Things: Recasting the Boundaries of the Mind*, Cambridge, UK, McDonald Institute for Archaeological Research

Marshack, A. (1972). *The Roots of Civilization: The Cognitive Beginnings of Man's First Art, Symbol and Notation*, New York, McGraw-Hill

Mithen, S. (1996). *The Prehistory of the Mind: The Cognitive Origins of Art, Religion and Science*, London, Thames and Hudson

Mithen, S. (2001). The Evolution of Imagination: An Archaeological Perspective. *Substance* 94/95: 48–50

Mithen, S. (2004). *After the Ice: A Global Human History 20,000–5000 BC*, London, Orion Books

Royal Museums Greenwich. (2012). *Royal Observatory Greenwich*, Northampton, UK, Belmont Press

Needham, J., Wang, L. and De Solla, D. (1986). *Heavenly Clockwork: The Great Astronomical Clocks of Medieval China* (2nd ed.), Cambridge, Massachusetts, Cambridge University Press

Orians, G. H. (2001). An Evolutionary Perspective on Aesthetics. *Bulletin of Psychology and the Arts* 2(1): 25–29

Renfrew, C. (1998). Mind and Matter: Cognitive Archaeology and External Symbolic Storage., in *Cognition and Material Culture: The Archaeology of Symbolic Storage*, C. Scarre (ed.), Cambridge, UK, McDonald Institute for Archaeological Research

Renfrew, C. (1973 [2011 digital]). *Before Civilization: The Radiocarbon Revolution and Prehistoric Europe*, Kindle Edition, Vintage Digital

Rosenberg, D. (2007). Marking Time. *Cabinet* 28 (Winter 2007/08), www.cabinetmagazine.org, accessed 7 July 2015

Ruggles, C. (ed.). (2015). *Handbook for Archaeoastronomy and Ethnoastronomy*, Berlin, Springer

——. (2011). *Archaeoastronomy and Ethnoastronomy: Building Bridges Between Cultures*, Cambridge, UK, Cambridge University Press

——. (2005). *Ancient Astronomy: An Encyclopaedia of Cosmologies and Myth*, Santa Barbara, US, ABC-CLIO

Spalinger, A. (1999). Egyptian Time, in K. Lippincott, *The Story of Time*, London, Merrell Holberton

Sommerville, W. B. (1972). Description of Foucault's Pendulum. *Quarterly Journal of the Royal Astronomical Society* 13: 40–62

Wallace, R. (1987). Introduction, in H. Blumenberg, *The Genesis of the Copernican World*, trans. R. Wallace, Cambridge, Massachusetts, MIT Press

Yiassemides, A. (2014). *Time and Timelessness: Temporality in the Theory of Carl Jung*, Hove, UK, Routledge

Zakbar, L. (1968). *The Study of the Ba Concept in Ancient Egyptian Texts*, Chicago, Illinois, University of Chicago Press

'Whatever is born, or done, in this moment of time, has the qualities of this moment in time.'

(Jung 1930: paras. 56–57)[1]

1 C. G. Jung. (1930 [1966]). Richard Wilhelm: In Memoriam, in CW, Vol. 15, *The Spirit in Man, Art and Literature*, London, Routledge & Kegan Paul.

INDEX

n refers to note.